Lecture Notes in Physics

W0037009

Edited by H. Araki, Kyoto, J. Ehlers, München, K. Hepp, Zürich
R. Kippenhahn, München, H. A. Weidenmüller, Heidelberg
and J. Zittartz, Köln
Managing Editor: W. Beiglböck

256

Dynamics of Wave Packets in Molecular and Nuclear Physics

Proceedings of the International Meeting
Held in Priorij Corsendonck, Belgium
July 2–4, 1985

Edited by
Jan Broeckhove, Luc Lathouwers and Piet van Leuven

Springer-Verlag Berlin Heidelberg GmbH

Editors

Jan Broeckhove
Luc Lathouwers
Piet van Leuven
Universiteit Antwerpen, Rijksuniversitair Centrum
Dienst Teoretische en Wiskundige Natuurkunde
Groenenborgerlaan 171, 2020 Antwerp, Belgium

ISBN 978-3-540-16772-3   ISBN 978-3-540-39847-9 (eBook)
DOI 10.1007/978-3-540-39847-9

2153/3140-543210

PREFACE

The international meeting on Dynamics of Wave Packets in Molecular and Nuclear
Physics was held at Priorij Corsendonck, Belgium, July 2-4, 1985. It was sponsored
by the NFWO (Nationaal Fonds voor Wetenschappelijk Onderzoek) of Belgium, the CECAM
(Centre Européen pour des Calculs Atomiques et Moléculaires), the RUCA (Rijksuniversitair Centrum Antwerpen, Belgium) and the Ministry of Education of Belgium. The
organizing committee consisted of J. Broeckhove (RUCA), L. Lathouwers (NFWO) and
P. Van Leuven (RUCA); the advisory committee consisted of E. Heller (University
of Washington, Seattle, USA), C. Joachain (University of Brussels, Belgium) and
C. Moser (CECAM, France).

 The meeting brought together 133 participants from the fields of nuclear physics,
molecular physics and dynamical systems, who contributed their expertise on the subject of the meeting, i.e. the time propagation of wave packets in quantum systems.
In these proceedings the written contributions submitted by the invited speakers
are reproduced. Other than retyping and reformatting they have not been subjected
to changes by the editors.

The Editors

J. BROEKCHOVE

L. LATHOUWERS

P. VAN LEUVEN

Dienst Teoretische en Wiskundige Natuurkunde

UNIVERSITEIT ANTWERPEN

Rijksuniversitair Centrum

Groenenborgerlaan 171, 2020 ANTWERP, BELGIUM

CONFERENCE PROGRAMME

Sessions	Speakers
Introduction	P. VAN LEUVEN
General	N. MANKOC-BORSTNIK
	N. MOISEYEV
	N. NIETO
	R. COALSON
Molecular	K. SINGER
	R. SKODJE
	C. LE FORESTIER
	J. BRICKMANN
	C. CERJAN
	H. MEYER
	L. LATHOUWERS
	J. REIMERS
Nuclear	B. GIRAUD
	R. ARVIEU
	R. DREIZLER
	J. BERGER
	J. GRIFFIN
	Y. HAHN
	K. KAN
	M. PLOSZAJCZAK
Chairman	J. BROECKHOVE

INTRODUCTION

Wave packets moving along classical paths in phase space have become a topic in those areas of quantum physics where a time-dependent description is appropriate. Such areas exist in both molecular and nuclear physics. Questions concerning semi-classical dynamics, phase-space organisation and quantisation procedures are basic to the use of wave packets. Unfortunately, molecular and nuclear physics are separated by a kind of barrier that makes the transfer of information quite difficult. This situation clearly entails the danger of duplication of effort.

The purpose of the metting on "Dynamics of Wave Packets in Molecular and Nuclear Physics" was to bring together specialists in molecular physics, nuclear physics and dynamical systems and to stimulate the interchange of ideas, concepts and techniques across the borderline separating these disciplines. At the same time this allowed each participant to assess the present status of research.

For the members of the organising committee, the meeting had a second more specific purpose, which was to explore the feasibility of a CECAM workshop on wave packet dynamics.

The meeting was very successful, due to the consistently high level of the contributions and the lively discussions generated by them. In this sense, the effort of organising the meeting was most fruitful.

The Organising Committee

TABLE OF CONTENTS

CLASSICAL PLANE TRAJECTORIES AND THEIR QUANTUM ANALOGS IN A BILLIARD AND TWO SPHEROIDAL CAVITIES

R. Arvieu

Institut des Sciences Nucléaires, Universite de Grenoble
53 Avenue des Martyrs, 38026 Grenoble Cedex, France.

1. Introduction

It is important to study simple systems in which the dynamics which results from classical mechanics can be compared to that coming out from quantum mechanics. In the hierarchy of dynamical systems [1]: integrable, pseudo-integrable and ergodic ones, the first ones are the easiest to deal with: in these systems the comparison between classical and quantum mechanics can be performed in the frame of the W.K.B. or E.B.K. semiclassical quanti- sation approximation. In all the domains where quantum mechanics is relevant, this method has known important successes. The aim of this talk is to present an example in which a single classical motion: that of a free particle in a domain with an elliptic boundary can be used to interpret three different quantum problems: the two dimensional elliptic membrane and the three dimensional prolate and oblate cavities. Our objective is threefold:

i) to study a simple problem in which the phase space is divided into two regions by a separatrix
ii) to emphasize the role of the Maslov indices in the semiclassical quantisation conditions and their ability to handle the dimension ana the symmetry of the system.
iii) to study the influence of the separatrix on the precision of the E.B.K. approximation.

2. Classical mechanics

The study of the motion of a free particle in a domain with elliptic boundary was already studied in detail by Keller and Rubinov [6]. A restricted presentation can be done if we perform the following steps:

2.1. CONSTANT OF MOTION

There exists a non-trivial constant of motion: the product $A = \vec{\ell}_1 . \vec{\ell}_2$ where $\vec{\ell}_1$ and $\vec{\ell}_2$ are the angular momentum of the particle with respect to each of the foci.

This constant of motion was found by Erikson and Hill [2] and by Helfrich [3] in problems with an elliptic symmetry, its role in our problem was underlined by Berry [4]. This constant allows to classify the trajectories into three subsets:

a) $A > 0$. The envelope of each trajectory is an ellipsis with the same foci as the boundary. If e is the eccentricity of the boundary, e_0 that of the envelope (caustic) we have by taking all positive values of A the domain of variation of e_0:

$$e \leq e_0 < 1 \quad . \tag{1}$$

b) $A < 0$. The envelope is now made by the two branches of an hyperbola homofocal to the contour. We have here by considering the set of negative values of A

$$e_0 > 1 \quad .$$

c) $A = 0$ (i.e. $e_0 = 1$). The trajectory always passes through each of the foci and plays the role of a **separatrix**. This constant of motion introduces potential barriers which will come out if we express the problem in proper coordinates.

2.2. SYSTEM OF COORDINATES

The symmetry of the boundary leads to very practical systems of coordinates. In two dimensions this system is the elliptical coordinates η and ξ defined by [5]

$$x = d \cosh \eta \cos \xi \tag{3}$$

$$y = f \sinh \eta \sin \xi \tag{4}$$

with
$$0 \leq \eta \leq \eta_1 \tag{5}$$

in which η_1 is related to the long semiaxis $R_>$ by

$$R_> = f \cosh \eta_1 \qquad (6)$$

2f being the focal distance. We define also

$$R_< = f \sinh \eta_1 \qquad (7)$$

and the deformation parameter

$$\mu = \frac{R_>}{R_<} = \tanh \eta_1 \quad . \qquad (8)$$

It is easily seen that the Hamilton Jacobi equations are separable in this system, this leads to the conjugate momenta p_η and p_ξ related to the energy k^2 (we use 2m=1) and to the separation constant E by

$$p_\eta^2 = k^2 f^2 \cosh^2 \eta - E \qquad (9)$$

$$p_\xi^2 = E - k^2 f^2 \cos^2 \xi \quad . \qquad (10)$$

We can easily show the following formula between E, A and kf

$$E = A + k^2 f^2 \quad . \qquad (11)$$

The turning points of the classical motion are found by $p_\eta (\eta_0) = 0$ (elliptic caustic, for A > 0) and by $p_\xi (\xi_0) = 0$ (hyperbolic caustic, A < 0). On the separatrix A = 0 we have:

$$p_\eta^2 = k^2 f^2 \sinh^2 \eta \qquad (12)$$

$$p_\xi^2 = k^2 f^2 \sin^2 \xi \quad . \qquad (13)$$

In three dimensions we must use the ellipsoidal prolate and oblate coordinates [5] η, ξ, ϕ (ϕ = angular coordinate around the axis of revolution). If we restrict ourselves to the motion with a zero projection of the angular momentum on the axis of symmetry the coordinate ϕ is ignorable. The system is still separable in η and ξ and the expressions (9) and (13) still hold. There is however a difference coming from the domain of variation of ξ.

$$0 \leq \xi < 2\pi \qquad \text{in 2 dimensions} \qquad (14)$$

there we can identify the points $\xi = 0$ with those with $\xi = 2\pi$ while

$$0 \leq \xi \leq \pi \qquad \text{in 3 dimensions} \qquad (15)$$

For a prolate ellipsoid the axis of revolution is characterised by the values $\xi = 0$ and $\xi = \pi$, for an oblate ellipsoid it is important to notice that the axis does not coincide with the focal line.

2.3. ACTION INTEGRALS

It is necessary to underline that the three problems under study are the same classically. Therefore they depend upon two action integrals I_η and I_ξ, each being defined as $1/2\pi$ times the integral of $\vec{p}.d\vec{q}$ along a complete cycle of motion of each variable. Those integrals were already given by Keller and Rubinov [6] in terms of elliptic integrals. The general structure of their results is for each action:

$$I = k \, f \, J(e,e_0) \qquad (15)$$

which reads: a trajectory in a billiard of eccentricity e and focal distance f with a caustic of eccentricity e_0, of energy k^2 has an action I (we use $\hbar = 1$) which depends on universal function J of the two eccentricities only.

A particularly important case is that of the action of the separatrix. Using (12), (13) and the definition of the action we find for $e_0 = 1$:

$$I_\eta = \frac{1}{\pi} \int_0^{\eta_1} k \, f \, \text{sh} \, \eta \, dy = \frac{1}{\pi} k(R_> - f) \tag{16}$$

$$I_\xi = \frac{1}{\pi} \int_0^\pi k \, f \, \sin \xi \, d\xi = \frac{2}{\pi} k \, f \quad . \tag{17}$$

This result may be understood in the following way: a trajectory with a given set of action I_η, I_ξ in a billiard the shape of which is changing adiabatically in time keeps, from Ehrenfest theorem, its actions constant when the shape is changing, this trajectory attains the sepratrix when the eccentricity of the billiard attains the values e_s given by the ratio of (16) and (17):

$$e_s = \frac{I_\xi}{2I_\eta + I_\xi} \quad . \tag{18}$$

Because the action integrals are the same for the two dimensional problem and the prolate cavity this formula holds equally well for the two cases. For the oblate cavity in the case of the elliptic caustic motion this equation is also valid. However it can be seen that the domain of variation of ξ in the case of the motion with hyperbolic caustic is exactly half of what it is for the same motion in the billiard. Therefore the corrosponding action I_ξ is exactly half of its values in two dimensions.

Formula (18) proofs that each trajectory evolves and attains the separatrix for a specific deformation. Moreover this deformation e_s, depends only on the ratio I_η/I_ξ, i.e. the way in which the motion is distributed between the η and the ξ variables.

2.4. NUMERICAL WORK

We have calculated the action and the energy i.e. the energy-action surface of a large variety of trajectories in billiard, prolate and oblate cavities as a function of the deformation parameter e or $\mu = \frac{R_>}{R_<}$. For the billiard problem the surface is kept constant, while the volume of the cavity is kept constant: a condition usually satisfied by nuclear matter. The trajectories which have action integrals which fulfill the semiclassical conditions have been calculated. Their energy spectrum will be discussed later on. We present in Fig. 1 the evolution of the eccentricity of the caustic e_0 as a function of μ for a prolate ellipsoid and for several orbits corresponding to deformed states. The labels $n\ell$ used here are connected to the quantum numbers of the spherical cavity (i.e. .$e_0 = 0$). It is seen that every orbit has a monotonous evolution: the starting point for $\mu = 1$ is $e_0 = 0$ (the caustic is a circle), then when μ increases, $0 < e_0 < 1$ (the caustic is an ellipsis), then $e_0 = 1$ (the separatrix !), finally e_0 gets always larger than 1 (the caustic is an hyperbola with an increasing eccentricity). Two rules are clearly seen:

a) The states with low ℓ perform their transition elliptic $\rightarrow$ hyperbolic faster than the states with high ℓ

b) In a parallel way the states with high n perform the transition faster than those with low n.

In order to understand this effect we must properly express in eq. (18) the quantised actions I_η and I_ξ for a prolate spheroid. We obtain for the critical eccentricity

$$e_s = \frac{\ell + \frac{1}{2}}{2(n + \frac{3}{4}) + (\ell + \frac{1}{2})} \qquad (19)$$

and for the critical value of the deformation parameter μ

$$\mu_s = \frac{2(2n+\ell+2)}{\sqrt{(4n+4\ell+5)(4n+3)}} \qquad . \qquad (20)$$

As an example let us consider the triplet of states: 3s, 2d, 1g which form a triplet degenerate for a spherical harmonic oscillator. We obtain for each of them:

$$3s \quad \mu_s = 1.033$$

$$2d \quad \mu_s = 1.100$$

$$1g \quad \mu_s = 1.512 \qquad .$$

For a moderate deformation $\mu \sim 1.2$ we understand that the states 3s and 2d have hyperbolic caustics while the state 1g is still elliptic, i.e., it has still a rather well defined value of its angular momentum.

It is interesting to note in Fig. 1 that the states with a very high ℓ like $\ell = 8$ (label 1k!) needs an extremely large value of μ ($\mu \approx 1.9$) to be on the separatrix.

3. Quantum mechanics

Although the classical trajectories are the same for the three problems under consideration the Laplacian contains terms [5] which are different in 2 and 3 dimensions. In addition the Laplacian is also different for the prolate and the oblate cases.

3.A. WAVE EQUATIONS

i) In two dimensions the wave function is simply written as:

$$\psi(\xi,\eta) = F(\xi)G(\eta) \tag{21}$$

with the following differential equations

$$\frac{d^2G}{d\eta^2} = (E - k^2f^2 \cosh^2 \eta)\, G \qquad \text{(Mathieu's equation)} \tag{22}$$

$$\frac{d^2F}{d\xi^2} = (k^2f^2 \cos^2\xi - E)\, F \qquad \text{(associated Mathieu's equation)} \tag{23}$$

ii) For a prolate cavity (modes of zero projection of the angular momentum on the axis of symmetry) we find using

$$\psi(\xi,\eta) = \frac{f(\xi)}{\sqrt{\sin \xi}} \frac{g(\eta)}{\sqrt{\sinh \eta}} \tag{24}$$

$$\frac{d^2g}{d\eta^2} = ((E + \tfrac{1}{4}) - k^2f^2 \cosh^2\eta + \frac{1}{4 \sinh^2\eta})\, g \tag{25}$$

$$\frac{d^2f}{d\xi^2} = (k^2f^2 \cos^2\xi - (E + \tfrac{1}{4}) - \frac{1}{4 \sin^2\xi})\, f \quad . \tag{26}$$

iii) For an oblate cavity we find with:

$$\psi(\eta,\xi) = \frac{g(\eta)}{\sqrt{\cosh\eta}} \frac{f(\xi)}{\sqrt{\sin\xi}} \tag{27}$$

$$\frac{d^2g}{d\eta^2} = ((E + \tfrac{1}{4}) - k^2 f^2 \cosh^2\eta + \frac{1}{4\cos^2 h\eta}) g \tag{28}$$

$$\frac{d^2f}{d\xi^2} = (k^2 f^2 \sin^2\xi - (E + \tfrac{1}{4}) - \frac{1}{4\sin^2\xi}) f \quad . \tag{29}$$

While the operators in the right hand side of (22) and (23) coincide with the expressions (12) and (13) of p_η^2 and p_ξ^2 respectively it is not the case in 3 dimensions, the main difference being due to terms which are singular for $\eta = 0$ or $\xi = 0$ and π. These terms can be treated approximately in the uniform approximation [7].

3.B. DIAGONALISATION

In order to provide an exact spectrum and a set of wave functions easy to manipulate we have calculated the energy levels in 2 and 3 dimensions by a diagonalisation in a basis of cylindrical and spherical Bessel functions respectively after having made a scale transformation of the coordinate axis. Using a basis large enough the spectrum of the elliptic billiard and that of the cavities have been calculated.

In Figure 2 we have plotted the part of the spectrum of the billiard which is obtained in the basis: $\{(1/\sqrt{2})(|n,m\rangle + |n,-m\rangle). \ m \neq 0; \ |n,0\rangle\}$ of cylindrical Bessel functions of $L_z = m =$ even (the label n is for the zeros of the Bessel functions). Four different basis uncoupled one from the other can be formed as well. We show here only the most symmetric ones. The levels are shown as a function of μ. For $\mu = 1$ the states are labelled by m_n.

In figure 3 we have plotted the energy levels of spheroidal cavities of constant volume. Only the levels of even parity are shown. They are also labelled using the zeros of the spherical unperturbed Bessel functions ℓ_n they correspond to for $\mu = 1$.

A few of the crossings have been studied carefully on Figures 2 and 3. Additional details and discussion will be done in an other publication [7].

3.C. SEMICLASSICAL QUANTISATION

i) W.K.B. (or E.B.K.) approximation

Both in 2 and 3 dimensions we must use the values (9) and (10) of p_η^2 and p_ξ^2 when we use the W.K.B. approximation to solve equations (22) to (29). In order to obtain this result it becomes necessary to neglect the additional terms in 3 dimensions. In the first approximation these terms are not completely neglected but as done with the semiclassical approximation of the Legendre polynomials [8] a proper phase of $\frac{\pi}{4}$ is introduced in the W.K.B. approximation. This prescription coincides with that of Keller and Rubinov [6] if the contour of integration in the action integral meets the axis of revolution an additional phase of $\frac{\pi}{4}$ should be introduced. On the other hand a uniform treatment of the wave functions in 3 dimensions [7] provides the same result. With the help of these short, but quite essential remarks, one finally gets the following table of quantisation.

The semiclassical quantisation lattices which are derived are presented in Figure 4 with an arbitrary position of the separatrix which is represented by a straight line (see eq. (18)) which rotates downward along the origin if μ increases.

It is quite striking to notice that the simplest case is the prolate one where the two lattices of quantisation coincide. In all the other cases the quantum cell allocated to each particle performs some motion when crossing the separatrix. For example the <u>cell of one particle in a symmetric state</u> in the billiard <u>slips towards a lower</u> I_η and <u>a higher</u> I_ξ while the antisymmetric cell is performing a motion which is exactly opposite to that one. The more complicated situation occurs for the oblate cavity in which there is a subtle factor of 2 change in I_ξ. Notice the similarity between the displacement of the even - ℓ cell of the cavity and that of the cell of the symmetric states of the billiard. On the other hand the odd - ℓ cell is comparable to the antisymmetric cell.

In conclusion one can estimate that the crossing of the separatrix leads really to important consequences in quantum mechanics because of this slipping of the unit cell.

	elliptic caustics		hyperbolic caustics	
	I_η	I_ξ	I_η	I_ξ
Elliptic billiard symmetric states	$n + \frac{3}{4}$	m	$n + \frac{1}{2}$	$m + \frac{1}{2}$
Elliptic billiard antisymmetric states ($m \neq 0$)	$n + \frac{3}{4}$	m	$n + 1$	$m - \frac{1}{2}$
Prolate cavity	$n + \frac{3}{4}$	$\ell + \frac{1}{2}$	$n + \frac{3}{4}$	$\ell + \frac{1}{2}$
Oblate cavity even parity	$n + \frac{3}{4}$	$\ell + \frac{1}{2}$	$n + \frac{1}{2}$	$\frac{\ell}{2} + \frac{1}{2}$
Oblate cavity odd parity ($\ell \neq 0$)	$n + \frac{3}{4}$	$\ell + \frac{1}{2}$	$n + 1$	$\frac{\ell}{2}$

Values of the action integrals for 2 and 3 dimensional problems.

ii) uniform approximation

Although we have not enough space to discuss this approximation properly we can say a few words about this possible improvement of the W.K.B. approximation. Indeed the potential barriers which are present both for the η and the ξ variables can be treated in a better way that what one does with constant Maslov's indices. For example the $\cos^2\xi$ and $\cosh^2\eta$ of the billiard problem can be developed up to second order and use can be made of the parabolic cylinder function to deduce the phase of the improved W.K.B. (E.B.K. 1). In this way the transition between the cells shown in Fig. 4 can be made in a continuous fashion and not a discontinuous one. This improvement, traced back to Ford and al. [9] has been made by us in each of our systems [7]. It is a necessity to understand in a better way the transition at the top of the barriers both for the η and for the ξ variables.

Numerical work

We have calculated the energy levels of the billiard and of the cavities using the lattices of quantisation shown above. Fig. 5 which corresponds to the billiard can be compared with Fig. 2 while Fig. 6 which corresponds to the cavities to Fig. 3. It is striking how the E.B.K. method is able to reproduce the details of the quantum spectrum for each deformation.

The orbitals associated to an elliptic caustic are represented by a dotted line in Fig. 6, those with a continuous line to the hyperbolic caustic. For each orbital the interval of the parameter μ corresponding to the elliptic caustic is the same as shown in Fig. 1 for the prolate case and to eq. (20). The downward concavity of the energy levels can be associated to the elliptic caustic. The change of concavity is associated to the change of caustic. These remarks are also true for the billiard problem. Notice that for the billiard and for an oblate cavity there is a failure of the semiclassical quantisation method to find the energy level at the top of the barrier but this can be performed with the help of the uniform approximation [7,9].

We compare in Fig. 7 the spectrum of a prolate cavity to that of the oblate cavity for the same value of the deformation μ = 2, the cavity having the same volume $\frac{4}{3}\pi R_0^3$. The semiclassical levels (S.C.) are plotted beside the quantum levels (Q). The states are still labelled by their quantum number n,ℓ for in the spherical situation they preserve these values in the semiclassical approximation. The states with odd - ℓ or even - ℓ lie in separated columns. The importance of the slipping of the unit cell in action space is now seen:

i) for the prolate case there is no slipping of the unit cell. The even ℓ states are then gently admixed with the odd ℓ states: the order for n = 0 is for example always s,p, d states etc... (It can be checked that this result holds independently on μ).

ii) for the oblate case the cell slips differently for the even - ℓ and the odd ℓ case. The odd parity states are then admixed in a more complicated way with the even ℓ states (i.e. the p level may cross the d or other states as well). Moreover the spectrum depends much on μ (which is not shown here). This result is the same as in the exact quantum spectrum which is also represented in Fig. 7. The discrepancies are generally very small ($\sim$ 0.5 %), the largest one are the absence of 6_1 for the prolate case and the presence of two levels 11_1, instead of a single one in quantum mechanics. However as said above these discrepancies can be removed by using the uniform approximation [7].

4. Conclusion

We have studied in this example in a systematic way the importance of having a phase space of two regions because of the presence of a simple constant of motion. The semiclassical vision of the phenomena is through a slipping of the unit cell in phase space which occurs only for the billiard problem and for the oblate cavity: the semiclassical orbits with hyperbolic orbits which correspond to symmetric states or to the even parity states tend to have a little more action in the ξ variable ("angular" motion) than in the η variable "radial" motion). This slipping in phase space produces a significant effect in the spectrum. It is doubtful that this simple and nice effect can be explained so simply outside the frame of the semiclassical theory.

This technique can be used for other problems in which a separatrix is present, for example for the modes of an oblate cavity with a non-zero projection of the angular momentum on the axis of symmetry or to those of a more realistic deformed potential [10]. It shows the full power of semiclassical theory to explain details that are hidden in a brute-force solution of the Schrödinger equation.

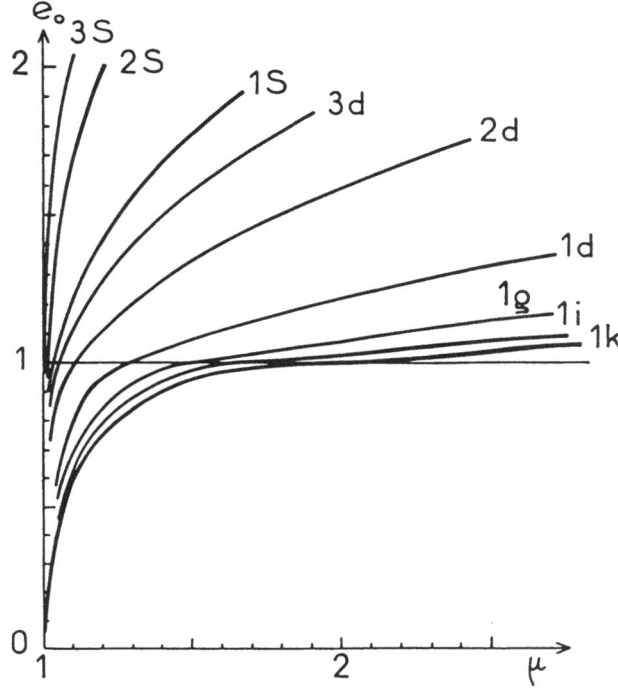

<u>Figure 1:</u> - The eccentricity e_0 of the caustic of the classical motion of trajectories of fixed actions I_η, I_ξ are studied as a function of μ ($=R_>/R_<$) for a prolate cavity. The states are labelled by the quantum numbers $n_r = n+1$. ℓ which are kept constant in the semi-classical approximation.

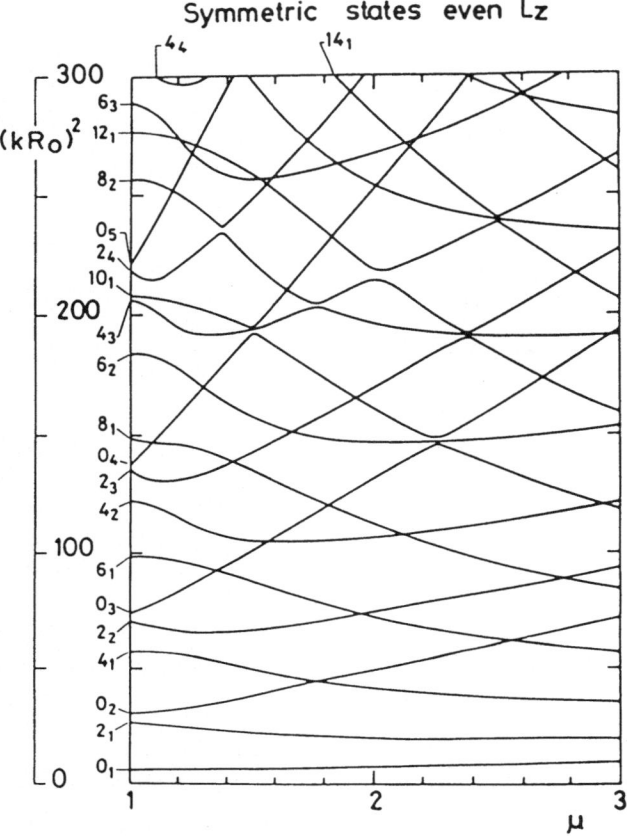

Figure 2: - The energy levels of the symmetric states of the two dimensional
billiard calculated by a diagonalisation method. States are
labelled by m_{n_r} where m is the projection of the angular momentum
(for $\mu = 1$) and n_r are the labels of the zeros of cylindrical Bessel
function.

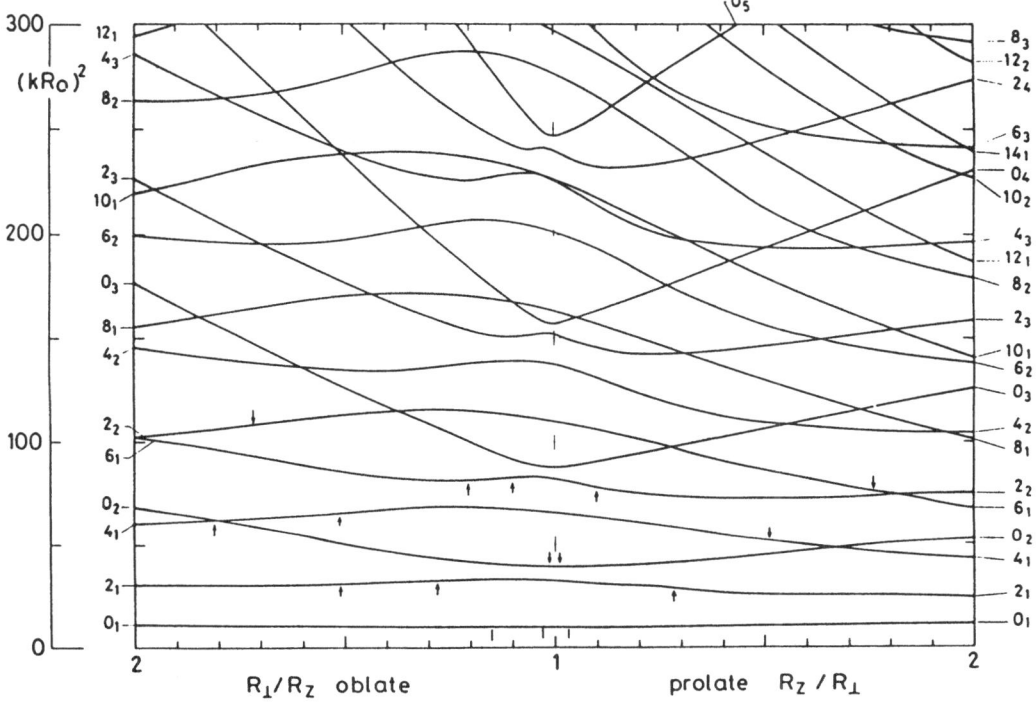

<u>Figure 3:</u> - The energy levels of the even parity states of the cavity as a
function of $\mu(=R_>/R_< = R_z/R_\perp$ (prolate case) $= R_\perp/R_z$ (oblate case)).
States are labelled by ℓ_{n_r}.

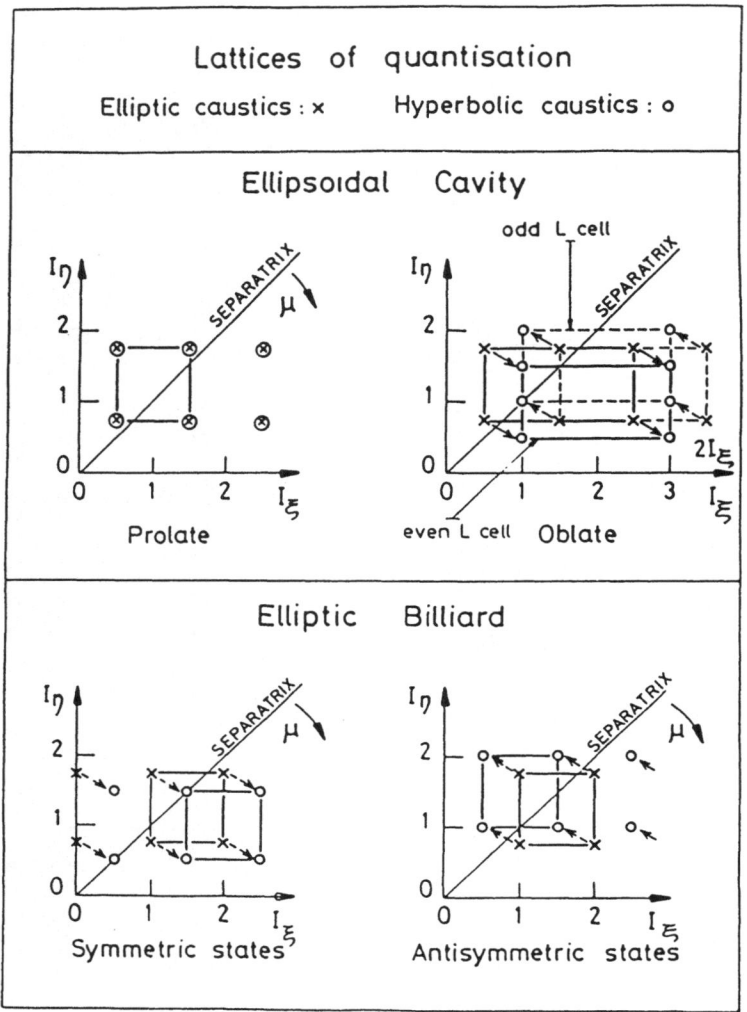

Figure 4: - Lattices of quantisation of the planar trajectories. The points above the separatrix correspond to trajectories which are quantized according to the hyperbolic caustic (circles), those below the separatrix to the trajectories quantised with an hyperbolic caustic (cross). When the deformation increases the separatrix, represented here by a line with an arbitrary orientation, rotates downward.

17

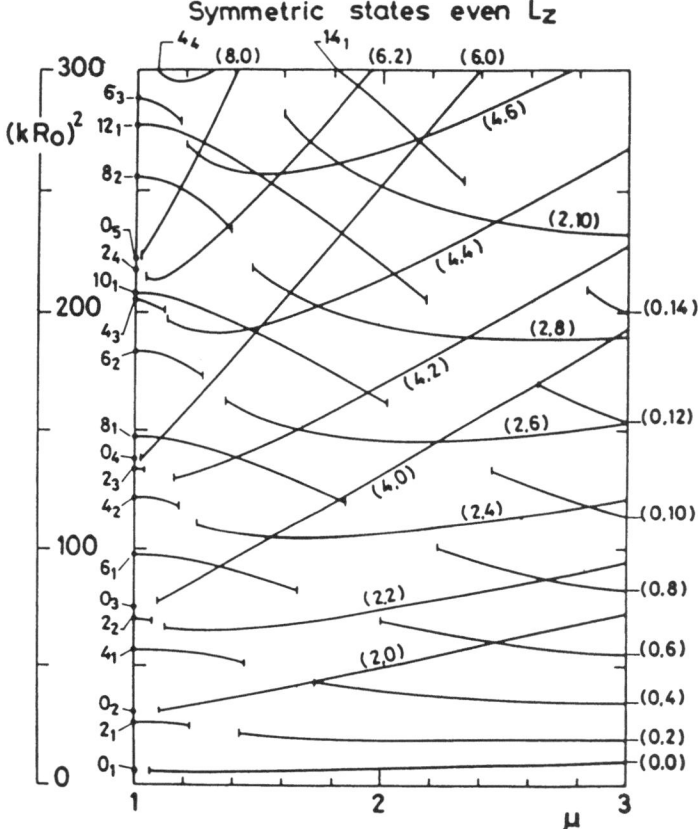

Figure 5: - The semiclassical spectrum of the elliptic billiard. We have
represented only the part of the spectrum associated to symmetric
states with even angular momentum. The labels on the left is m_{n_r}
where m is the angular momentum and n_r is the label of the zero of
the cylindrical Bessel functions (n_r = n+1). The left part of the
curve is associated to an elliptic caustic, the curve on the right
to the hyperbolic one. The labels in the right are simply (n,m).
The separatrix is seen by an interruption between the two branches.
To be compared to Fig. 2.

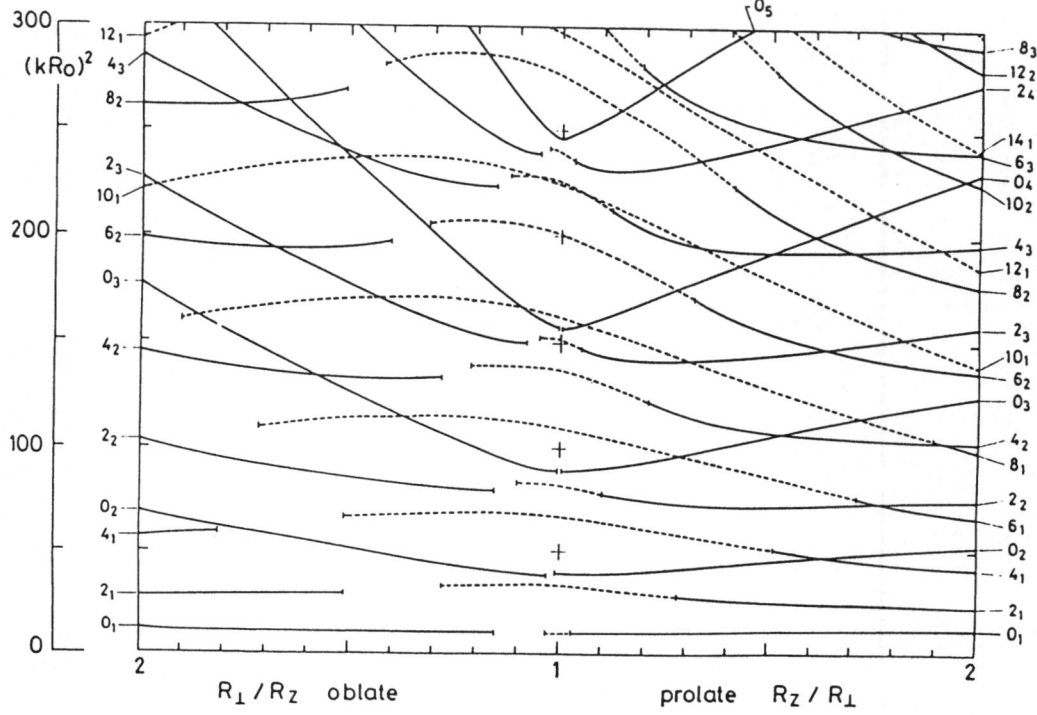

Figure 6: - The semiclassical spectrum of the even parity states of the cavity. States are labelled by ℓ_n (where $n_r = n+1$). The dotted line corresponds to an elliptic caustic, the continuous line to the hyperbolic caustic. The transition is continuous for the prolate case (see Fig. 4), but discontinuous for the oblate one. To be compared to Fig. 3.

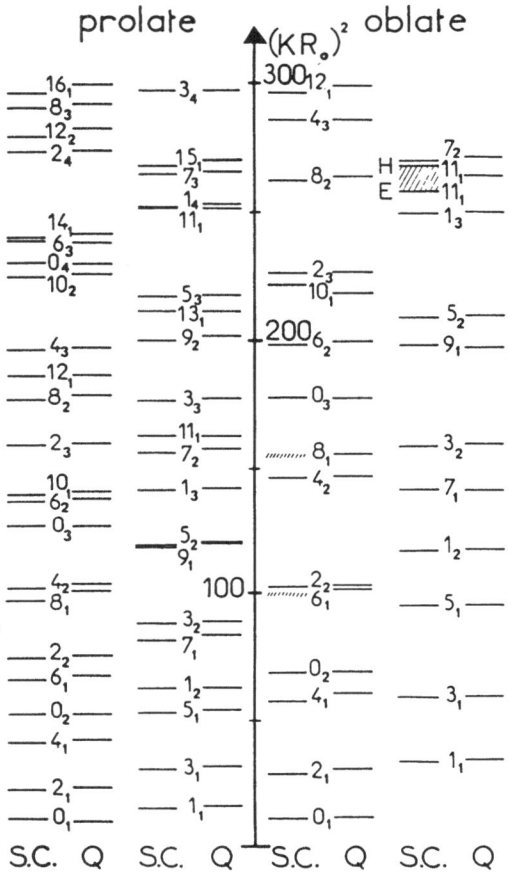

Figure 7: - The effect of the different slipping of the phase space cell (I_n, I_ζ) for oblate and prolate spheroid of same volume and deformation $(\mu=R_>/R_< = 2)$. The semiclassical spectrum (S.C.) is presented beside the exact quantum spectrum (Q). The slipping produces: a less dense spectrum for the oblate case and a complex admixture of even and odd parity states. The 6_1 state is missing in the semi-classical approximation, on the other hand this method provides two condidates for 11_1 with different caustics (marked respectively H and E).

References

1) The litterature on the study of dynamical systems has grown tremendously in the last few years. A comprehensive survey is presented in several review articles by M.V. Berry, like: "Topics in nonlinear dynamics", A.I.P. Conference proceedings, N.Y. (1978) 16; Les Houches, Session XXXVI (1981) North Holland, 171.

2) H.A. Erikson and E.L. Hill, Phys. Rev. 75 (1949) 29

3) K. Helfrich, Teoret. Chem. Acta (Berlin) 24 (1972) 271

4) M.V. Berry, Eur. J. Phys. 2 (1981) 91

5) P. Moon and D.E. Spencer, Field theory handbook, Springer Verlag (1961)

6) J.B. Keller and S.I. Rubinov, Ann. of Phys. 9 (1960) 24; 10 (1960) 303

7) R. Arvieu and Y. Ayant, to be published.

8) L. Landau and E. Lifshitz, Quantum Mechanics (MIR 1974) ch. VII - § - 49

9) K.W. Ford, D.L Hill, M. Wakano and J.A. Wheeler, Ann. of Phys. 7 (1959) 239

10) F. Brut and J. Carbonell, private communication.

QUANTUM DYNAMICS OF WAVEPACKETS ON TWO-DIMENSIONAL POTENTIAL ENERGY SURFACES GOVERNING NUCLEAR FISSION

J.F. Berger

Service de Physique Neutronique et Nucléaire, Centre d'Etudes de Bruyères-le-Châtel, B.P. n° 12, 91680 Bruyères-le-Châtel, France

The dynamics of nuclear systems undergoing a large re-arrangement of nucleons, such as actinide fission, is generally described with the help of collective coordinates. This procedure, aimed at reducing the very large number of degrees of freedom otherwise needed, is expected to be particularly useful at low energy (for instance, in the vicinity of the fission threshold), since the coherent motion of all the nucleons is much slower than their internal motion in the field of the fissioning nucleus. The internal degrees of freedom are therefore almost completely relaxed and the dynamics of the phenomenon is mainly governed by the collective properties of the system.

Most studies developed along this line in the past years introduce the collective coordinates in a quasi-classical picture of collective motion based on the liquid-drop model [1-3]. The influence of the internal structure-shells, pairing correlations - is generally incorporated in a phenomenological way. One then derives a potential energy surface and, in some cases, an inertia tensor that can be used to build a classical collective hamiltonian. Basic features of the fission process could be reproduced with this kind of approach, notably, fission isomerism and asymmetric mass division. However, the applicability of such phenomenological models to the details of fission dynamics, especially at large deformation, is doubtful. In fact, nuclear collective properties, in particular the collective inertia, are very sensitive to internal effects such as pairing correlations, that cannot be accurately estimated on a phenomenological basis. Further, quantum mechanical aspects must be reincorporated in the description, since collective variable fluctuations and barrier penetration are expected to play an important role in low energy fission. Semi-classical approaches or functional integral techniques are not easy to use when more than one collective variable comes into play - the general case, while a full treatment requires the collective hamiltonian to be quantized, which cannot be done in a unique way.

In this talk, we present an approach to the fission process that is free
of these limitations since it is a completely microscopic and fully quantum
one. It is based on the Hartree-Fock-Bogolyubov method - that extension of
the Hartree-Fock procedure including a self-consistent microscopic treatment
of pairing correlations - with several external constraining fields. The latter
are used to probe the static collective properties of the system, in particular
the kind of collective modes that have to be considered. The dynamical aspects
of the phenomenon are studied in a time-dependent formalism starting with
a dynamic state of Generator Coordinate form. Using the Gaussian Overlap
Approximation, a quantum collective hamiltonian taking into account zero-point
energy fluctuations can be derived. This leads to a time-dependent Schrödinger-
like equation giving the evolution of a collective wave-function containing
all the dynamical information.

For the sake of completeness, we first briefly recall the principle of the
microscopic approach outlined above (a more detailed account can be found in
Ref. [4]). Then, we concentrate on the dynamical method and, in particular, on
the techniques employed to determine the propagation of collective wavepackets
on the potential energy surfaces with a coordinate-dependent inertia tensor.
Although the method can be applied for the whole phenomenon, we shall focus
on the behaviour of the nucleus beyond the second saddle point and on the
scission process, since these aspects of fission are poorly known.
At the same time, the observable fragment characteristics (e.g. mass, charge,
kinetic and excitation energy distributions) closely depend on the dynamics
at large deformation.

1. The microccopic method

As mentioned above, the collective aspects of fission are probed in our
approach by means of the Hartree-Fock-Bogolyubov (HFB) method with several
external constraining fields. The external fields are taken proportional to
some low order multipolar operators $\hat{Q}_{\ell m}$ in order to test different types of
nuclear deformation. These fields are added to the microscopic nuclear
hamiltonian $\hat{H}$ and the HFB variational principle:

$$\delta < \phi_{\{q\}} | \hat{H} - \sum_{\ell} \lambda_{\ell m} \hat{Q}_{\ell m} - \lambda_N \hat{N} - \lambda_Z \hat{Z} | \phi_{\{q\}} > = 0$$

is solved with the constraint conditions:

$$\langle \phi_{\{q\}} | \hat{Q}_{\ell m} | \phi_{\{q\}} \rangle = q_{\ell m} \quad , \quad \{q_{\ell m}\} \equiv \{q\} \quad . \tag{1}$$

This gives the independent quasi-particle states $|\phi_{\{q\}}\rangle$, from which all quantities pertinent to fission (e.g. deformation and pairing energies, nucleon distributions) can be derived. The main advantage of this approach - which requires a considerable numerical effort - is to include in a self-consistent way the effects of the rearrangement of the average and pairing fields with deformation, while the only ingredient is the nucleon-nucleon interaction. In what concerns the latter, our calculations are performed with the finite range force D1 [5]. It is particularly suitable for fission studies since its pairing properties have been carefully adjusted [4].

The set of chosen multipolar deformations $q_{\ell m}$ can clearly be considered as collective coordinates and a dynamical approach of the Adiabatic Time-Dependent Hartree-Fock (ATDHF) type [6] could be used. However, in order to keep a full quantum-mechanical description, the dynamics state of the fissioning nucleus is expressed as in the Generator Coordinate (GC) theory [7]:

$$|\psi(t)\rangle = \int d_{\{q\}} \, \chi_{\{q\}}(t) \, |\phi_{\{q\}}\rangle \quad . \tag{2}$$

As is well known, a time-dependent variational principle applied to (2) leads to an integral equation of Hill-Wheeler type for the unknown $\chi_{\{q\}}(t)$. By use of the Gaussian Overlap Approximation (GOA) [8], this integral equation can be transformed into a time-dependent Schrödinger-like equation

$$\hat{\mathcal{H}} \, \bar{\chi}_{\{q\}}(t) = i\hbar \, \frac{\partial \bar{\chi}_{\{q\}}(t)}{\partial t} \quad . \tag{3}$$

Here $\bar{\chi}_{\{q\}}(t)$ is a collective wave-function related to $\chi_{\{q\}}$ by means of an overlap kernel and normalized by

$$\int d_{\{q\}} |\bar{\chi}_{\{q\}}(t)|^2 = 1 \quad . \tag{4}$$

$\hat{\mathcal{H}}$ is a collective hamiltonian which appears directly in quantized form:

$$\hat{\mathcal{X}} = - \frac{\hbar^2}{2} \sum_{j,k} \frac{\partial}{\partial q_j} (M_{\{q\}}^{-1})_{jk} \frac{\partial}{\partial q_k} + \langle \phi_{\{q\}} | \hat{H} | \phi_{\{q\}} \rangle - \Delta E_{\{q\}} \quad . \tag{5}$$

In this expression, the indices j and k stand for the multipolar labels (ℓ,m) of the chosen constrained collective parameters, $M_{\{q\}}$ is the matrix of the collective inertia tensor, $\langle \phi_{\{q\}} | \hat{H} | \phi_{\{q\}} \rangle$ the HFB deformation energy and $\Delta E_{\{q\}}$ is a quantum correction to the potential energy coming from the fluctuations of the collective variables in the states $| \phi_{\{q\}} \rangle$. This correction can be easily expressed in terms of the HFB quasi-particle states [8]. As for $M_{\{q\}}$ it has been shown that one must take the expression given by the ATDHF theory [9]. The HFB procedure then provides all the elements necessary to solve the propagation equation (3) and, consequently, to achieve a completely microscopic and quantum description of fission dynamics.

To conclude this section, let us note that the kinetic energy term does not appear in (5) in the usual Pauli prescription form. However, it is easy to show that the two expressions differ only by a scalar term containing derivatives of the inertia tensor. One could therefore use Pauli form for the kinetic energy in (5) provided $\Delta E_{\{q\}}$ is appropriately modified.

2. Fission dynamics at large deformation

2.1. COLLECTIVE COORDINATES

In the study of actinide fission, two external constraining fields were first used: $\hat{Q}_{20}$ governing the overall deformation of the nucleus and $\hat{Q}_{30}$ its left-right asymmetry. However, at large deformations, two distinct families of self-consistent solutions appeared. One described the stretching of a unique nucleus, while the other corresponded to two well-separated fragments. Such a situation indicated that another collective variable that would permit continuous passage from one family to the other was missing in our approach. This collective variable could clearly be associated with the scission process. Consequently, we introduced in the HFB procedure an additional constraint proportional to the hexadecapole moment $\hat{Q}_{40}$. This choice was natural since $\hat{Q}_{40}$ can control the necking-in of the nuclear density distribution. Self consistent calculations performed with this new constraint indeed showed that a continuous set of HFB solutions could be obtained. These solutions included both nuclear configurations before and after scission. Thus, the microscopic analysis of the collective modes tells us that, in addition to elongation and mass-asymmetry, a necking-in collective coordinate must be taken into account at large deformation.

The necking-in coordinate appears to play a crucial role in the dynamics of fission. This can be viewed from the structure of the Potential Energy Surface (PES) obtained when this collective coordinate is included. Fig. 1 displays the large deformation PES of ^{240}Pu as a function of $q_2 = \langle\hat{Q}_{20}\rangle$ and $q_4 = \langle\hat{Q}_{40}\rangle$, that is

$$V(q_2,q_4) = \langle\phi_{q_2 q_4}|\hat{H}|\phi_{q_2 q_4}\rangle - \Delta E(q_2,q_4) \quad . \tag{6}$$

The octupolar operator $\hat{Q}_{30}$ has not been constrained in this case. Consequently, the left-right asymmetry of the system takes values that minimize the deformation energy for given q_2 and q_4 (it remains in the vicinity of the 106/134 fragmentation over most of the PES). This surface shows two valleys V_1 and V_2 separated by a barrier whose height decreases with q_2-deformation and vanishes around $q_2 = 370b$. The V_1 valley describes nuclear configuration before scission, while the nucleus is completely fragmented in V_2. The shaded area near the top of the barrier indicates when two tangent fragments are observed. It therefore appears that scission may occur at different deformations for given initial conditions. Most of scissions should occur near $q_2 = 370b$ where the barrier vanishes. However, they may also take place at lower deformations. In this case the nascent fragments are formed closer to each other and have comparatively little deformation energy. Consequently, these "compact scissions" would lead to high fragment kinetic energy and low internal excitation energy. These compact scissions are very likely related to the so-called "cold fission" events observed experimentally in actinide nuclei [10-12].

In order to obtain a quantitative estimate of the relative probability of compact scissions, a dynamical calculation for the collective evolution on the $V(q_2,q_4)$ PES of Fig. 1 has been made. The dynamical formalism developed in the preceeding section has been used so as to take into account dynamical effects (e.g. variation of the inertia, influence of wavepacket velocity and spreading) that are expected to play an important role in this study. The inertia tensor $M(q_2,q_4)$ was derived from the cranking formula. Numerical results indicate that M is practically constant along lines parallel to the V_1 and V_2 valleys. As a consequence, the inertia is a function $M(x)$ of the orthogonal distance x from the bottom of the V_1 valley. The matrix elements of $M(x)$ are plotted in Fig.2. One can see that the cross term M_{24} is not negligible especially in V_1 (x=0) and in V_2 (x≈40) while its absolute value decreases near the top of the barrier (x≈12) and in the vicinity of the scission line (x≈22).

We now present the method used to solve the propagation equation (3) in the case of two variables -$\{q\} = (q_2, q_4)$ - and with $\hat{\mathscr{H}}$ given by (5).

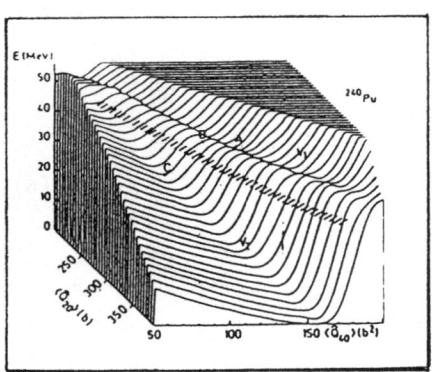

Fig. 1. Potential energy surface $\overline{V}(q_2, q_4)$ of ^{240}Pu at large deformation for the mass-asymmetry that minimizes the deformation energy. The hatched area represents the scission region.

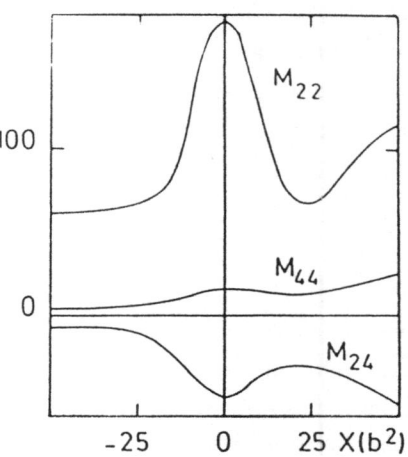

Fig. 2. Matrix elements of the collective inertia tensor governing q_2-q_4 dynamics. X is defined in the text. M_{jk} is in units of $\hbar^2$ $A^{(j+k+6)/3}$ MeV^{-1} fm^{-j-k} with $A = 240$.

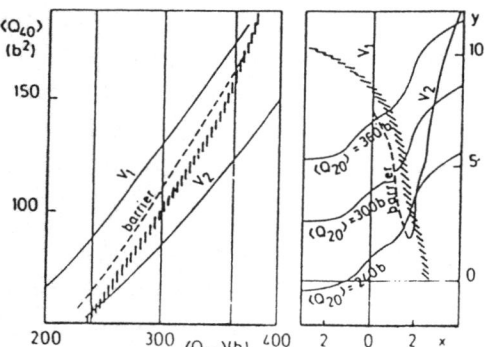

Fig. 3. Effect of the change of collective variables $(q_2, q_4) \rightarrow (x, y)$ on characteristic lines of the q_2-q_4 potential energy surface. The hatched areas represent the scission region.

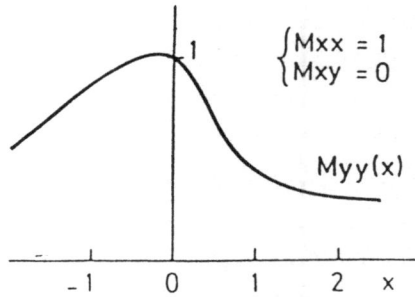

Fig. 4. Matrix elements of the collective inertia tensor in the x-y variables.

2.2. METHOD OF SOLUTION OF THE DYNAMICAL EQUATIONS

The first step of our method is to introduce new collective variables (x,y) in order to diagonalize the kinetic energy term of the hamiltonian. It is well known that such a transformation always exists in two dimensions if the determinant of M does not change sign (which is our case). One of the new variables, x say, can be chosen such such that the bottom of V_1 is the x=0 line. Moreover, since M depends on q_2 and q_4 only through the particular linear combination X, the transformed diagonal term M_{xx} can be taken equal to 1, while M_{yy} depends on x. The effect of this change of collective variables is displayed in Fig. 3. The dependence of M_{yy} on x is shown in Fig. 4. We note that this coordinate transformation defines pseudo-orthogonal collective modes: a scission mode along the x-variable and an elongation mode along y.

With these coordinates, the hamiltonian (5) is close to the usual Schrödinger hamiltonian:

$$\hat{\mathscr{H}} = - \frac{h^2}{2} \frac{\partial^2}{\partial x^2} - \frac{h^2}{2} \frac{1}{M_{yy}(x)} \frac{\partial^2}{\partial y^2} + V(x,y) \quad . \tag{7}$$

The solution of the time-dependent propagation equation is obtained by timesteps Δt:

$$\bar{\chi}(x,y,t+\Delta t) = \exp\left(\frac{\Delta t}{i\hbar} \hat{\mathscr{H}}\right) \bar{\chi}(x,y,t) \tag{8}$$

starting from an initial condition $\bar{\chi}_0(x,y)$. The latter has been chosen in order to simulate a fission event emerging at the beginning of the V_1 valley under the fission threshold. It is written as

$$\bar{\chi}_0(x,y) = \phi_0(x) \times G(y)$$

where $\phi_0(x)$ is the lowest state in the potential well $V(x,y=0)$ and $G(y)$ a gaussian wave-packet centred at y=0 with zero mean velocity.

The equation (8) has been approximated by finite differences on a rectangular mesh defined in the part of the x-y plane represented in Fig. 3. It therefore reduces to a matrix equation. In order to evaluate the matrix of the propagator $\exp\left(\frac{\Delta t}{i\hbar} \hat{\mathscr{H}}\right)$, the kinetic energy term $\hat{T}_x$ containing the x-variable is separated from the remaining of the hamiltonian

$$\hat{H}_y(x) = -\frac{\hbar^2}{z}\frac{1}{M_{yy}(x)}\frac{\partial^2}{\partial y^2} + V(x,y) \qquad (9)$$

and the approximate expression is used

$$\exp\frac{\Delta t}{i\hbar}\mathcal{H} = \frac{1}{2}\{\exp\frac{\Delta t}{i\hbar}\hat{T}_x \cdot \exp\frac{\Delta t}{i\hbar}\hat{H}_y(x) + \exp\frac{\Delta t}{i\hbar}\hat{H}_y(x)\cdot\exp\frac{\Delta t}{i\hbar}\hat{T}_x\}$$

which is true through order $(\Delta t)^3$. The matrices of the partial propagators are easy to compute since $\hat{T}_x$ depends only on x and $H_y(x)$ is diagonal in the x variable. The matrices of $\hat{T}_x$ and $H_y(x)$ are diagonalized (the latter for each point of the x-mesh) so as to express the exponentials exactly. Good convergence of the results is obtained with a 61 × 141 (x-y) rectangular mesh and a time step $\Delta t = 10^{-23}$ s (the characteristic time of collective evolution is a few 10^{-21} s).

The main difficulty encountered in the numerical determination of $\bar{\chi}(x,y,t)$ concerns the reflections of the wave-function at the boundary of the (x-y) domain used. In fact, the initial wave-packet tends to spread over the entire domain long before its centre of mass crosses the scission region. Reflected components then mix with the original wave-packet leading to a very oscillatory wave-function that ultimately fills the whole domain. Various types of time-dependent boundary conditions were tried and tested in a flat potential model with no satisfactory results. Also, attempts to evolve the wave-function outside the domain by means of approximate equations led to partial reflections because the outside and inside parts of the wave-function do not match each other with sufficient accuracy. In view of these difficulties, a completely different technique has been employed: the wave function is progressively absorbed within a narrow region along the boundary of the domain. More precisely, after each time-step Δt, the wave function $\bar{\chi}(x,y,t+\Delta t)$ is weighted by a factor $\exp(-A(x))$ in the region $x_2 - d_x < x < x_2$ close to the $x = x_2$ right boundary. Similar weight factors $\exp(-A(y))$ are applied in the vicinity of the $y = y_1$ and $y = y_2$ boundaries. (No absorption is necessary near the $x = x_1$ left boundary, since $V(x,y)$ strongly increases for negative values of x). Saxon-Woods forms have been taken for $A(x)$ and $A(y)$. For example, in the neighbourhood of $x = x_2$, $A(x)$ is given by

$$A(x) = A_0/(1+\exp -\alpha(x-x_2+dx/2)) \qquad .$$

Numerical tests showed that the spurious reflections are avoided when the absorption of the wave-function is made very progressive. To this purpose A_0 and α have to be chosen as small as possible. In this case the widths d_x and d_y of the absorption regions must be sufficiently large. The present calculation has been performed with the following values: $A_0 = 7$ MeV, $\alpha = 10$, $d_x = .6$, $d_y = .5$ near $y = y_1$ and $d_y = 0$ near $y = y_2$. We observed that a notable change of these parameters, for instance A_0 and α increased by a factor of two, affects the behaviour of the wave function essentially in the vicinity of the absorption regions. In fact, the perturbations cause by partial reflections do not propagate back into the scission region, even for large times. With larger values of A_0 and α, however (e.g. $A_0 = 50$ MeV or $\alpha = 100.$) the wave-function is progressively spoiled over all the (x-y) domain.

It can be noted that the technique described above consists in replacing the finite dimensional evolution operator $\exp \frac{\Delta t}{i\hbar} \mathcal{H}$ by

$$\exp \frac{\Delta t}{i\hbar} \mathcal{H} \; . \; \exp (-A) \quad .$$

The absorption part can therefore be viewed as coming from an imaginary potential term added to the collective hamiltonian $\hat{\mathcal{H}}$ in the neighbourhood of the domain boundary. In particular, if one neglects the $[\hat{\mathcal{H}}, A]$ commutator, this imaginary part simply reduces to $-i\hbar A/\Delta t$.

2.3. RESULTS

The main results obtained with the above dynamical model are summarized in Fig. 5. This figure displays the flux that passed from time $t = 0$ to times $t = 1,2,\ldots,7$, through successive length elements of the scission line. The abscissa gives the values of q_2-deformation along this line. Times are in units of 10^{-21}s. Two conclusions can be drawn from the behaviour of the curves:

i) Most scission events occur in the deformation region where the barrier between the V_1 and V_2 valleys vanishes (i.e. $q_2 \sim 370$ b). However, compact scissions are also observed, though with lower probability. In particular a small proportion of scissions takes place at deformations less than $q_2 = 280$ b. The HFB results show that the associated fragments have deformation energies smaller than 3 MeV and kinetic energies higher than 208 MeV. Therefore, they clearly are "cold fission" events. Their relative yield is about 4.10^{-4} which is consistent with experimental measurements in ^{240}Pu [11]. It can be noted that the 106/134 fragmentation which appears as the most probable in our approach, also gives the highest yield in cold fission experimental data [10-12].

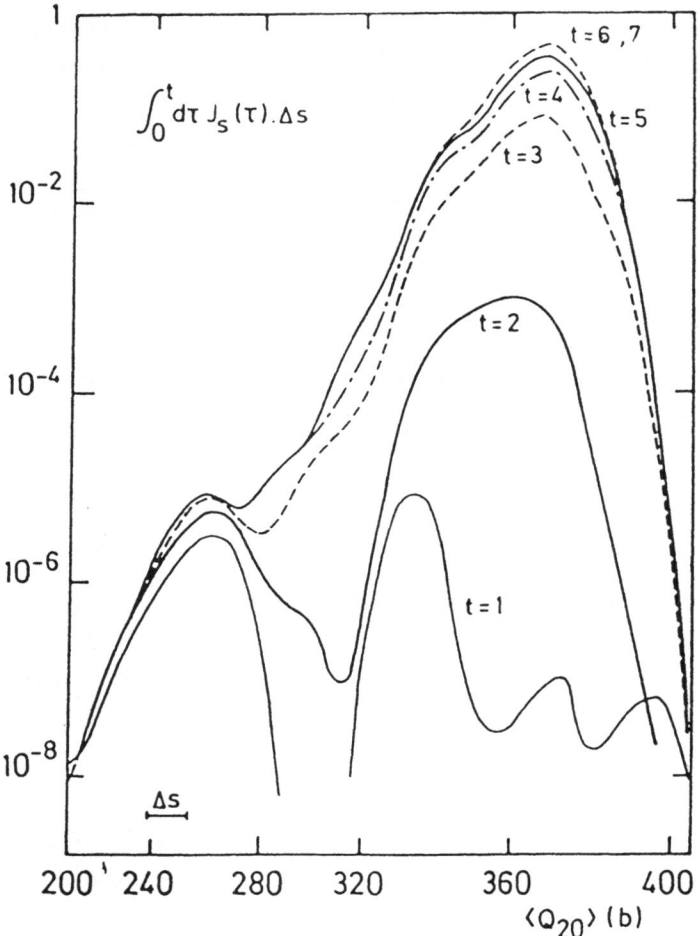

$$\int_0^t d\tau \, J_s(\tau) . \Delta s$$

Fig. 5: Cumulated flux through successive length elements along the scission line at different times. Times are in units of 10^{-21} s.

ii) The characteristic time for the most probable scissions is about 3×10^{-21} s, which is the order of magnitude of the time of evolution between saddle and scission in actinides. Such a large figure may look surprising since no dissipative term is included in our dynamical description. Indeed, more phenomenological models have been compelled to introduce some form of viscosity in order to reproduce characteristic times of this order of magnitude [13].

In fact, a careful analysis of the evolution of the collective wave function shows that the above results are essentially explained by the effects of quantum wave-packet propagation in a two-dimensional collective space:

i) At the beginning of the collective evolution, the state rapidly spreads in the V_1 valley. Meanwhile its group velocity is still small, because of the large value of the M_{yy} inertia in V_1. Consequently, the low deformation tail of the wave-packet is practically at rest in the upper part of V_1 and has enough time to tunnel through the barrier into V_2. This phenomenon explains why compact scissions can occur with a finite probability in spite of large values (> 4 MeV) of the barrier height.

ii) During the descent starting from the upper part of V_1, an increasing proportion of the available potential energy is stored in the x-mode, orthogonal to the stretching mode. This effect, essentially due to the particular structure of the $V(q_2,q_4)$ PES, contributes to slow down the motion of the wave-packet along the V_1 valley. As a consequence, the probability for compact scission is enhanced while the characteristic time for the collective evolution in the V_1 valley is much longer than expected. It can be said that collective dynamics in the x-direction plays the role of a dissipation mode with respects to the dynamics in the stretching direction. In fact, no other dissipation effect is needed in order to explain fission characteristic times of a few 10^{-21}s. It is important to note that a slow collective velocity is consistent with the adiabatic hypothesis upon which our dynamical description rests.

In order to illustrate these remarks, we give the energy repartition of a fission event starting at t=0 from the beginning of the V_1 valley ($q_2 \sim 220b$) with 4 MeV excitation energy in ^{240}Pu. Since this energy is under the fission threshold, the initial kinetic energy in the stretching mode T_y is zero. At $t \sim 2.10^{-21}$s, the centre of the wave-packets gets close to $q_2 = 280b$ in the V_1 valley and the potential energy gained during the descent is 4.5 MeV. However, only 2 MeV appear in the T_y kinetic energy; the remaining 1.5 MeV is stored as collective vibration energy in the x-mode. Similarly, when the wave-packet centre reaches the most probable scission region ($q_2 \sim 370b$) at $t \sim 3.10^{-21}$s, the potential energy has decreased by 11.5 MeV. At the same time, the total kinetic energy is only 9.4 MeV. The 2.1 MeV difference has been transformed into x-vibration energy. Just after scission, this energy will appear essentially as collective excitations in the fission fragments.

3. Conclusion

Three kinds of conclusion can be drawn from the present study of low energy fission dynamics at large deformation:

i) a microscopic analysis of nuclear collective properties by means of the Hartree-Fock-Bogolyubov method with external fiels gives information on the type and the number of the collective variables which are relevant to the process. Further, a description in terms of at least two collective variables is necessary at large deformation in actinides in order to obtain a set of deformed self-consistent states which is continuous.

ii) low energy collective dynamics can be derived in the framework of a microscopic quantum model where the collective hamiltonian appears directly in quantized form, with the appropriate zero-point energy corrections. The collective inertia which is generally non-diagonal and coordinate dependent must be carefully taken into account since it strongly influences the coupling between collective modes. When solving numerically the time-dependent Schrödinger-like equation for the collective wave-function, spurious reflections at the boundary of the finite domain used have to be avoided. This can be done by means of very simple absorption technique.

iii) quantum effects play an important role in collective dynamics. In particular, wave-packet spreading and the cross-fluctuations between two collective modes strongly affect the behaviour of the tail of the wave-functions and the characteristic times for collective evolution. In this study, quantum effects appear essential to explain the occurence of compact scissions and, consequently, the cold fission phenomenon. Also, the quantum dynamical description demonstrates that collective motion in fission is slow, which a posteriori justifies the adiabatic hypothesis.

It is a pleasure to thank Dr. K. Moser for his constant interest in this work and for his assistance in solving certain numerical problems. We enjoyed stimulating discussions on the content of this paper with R. Walsh and are grateful for many valuable comments and suggestions.

References

1) M. Brack, J. Damgaard, A.S. Jensen, H.C. Pauli, V.M. Strutinsky and C.Y. Wong, Rev. Mod. Phys. 44 (1972) 320.

2) P.Möller and J.R. Nix, Physics and Chemistry of Fission (IAEA, Vienna, 1974)

3) R.W. Hasse, Nucl. Phys. A128 (1969) 609.

4) J.F. Berger, M. Girod and D. Gogny, Nucl. Phys. A428 (1984) 23c.

5) J. Decharge and D. Gogny, Phys. Rev. C21 (1980) 1568.

6) M. Baranger and M. Veneroni, Ann. Phys. 114 (1978) 123.

7) D.L. Hill and J.A. Wheeler, Phys. Rev. 89 (1953) 1102.
   J.J. Griffin and J.A. Wheeler, Phys. Rev. 108 (1957) 311.

8) B. Giraud and B. Grammaticos, Nucl. Phys. A233 (1974) 373; A255 (1975) 141.
   M. Girod and B. Grammaticos, Nucl. Phys. A330 (1979) 40.

9) F.M.H. Villars, Nucl. Phys. A420 (1984) 61.

10) C. Signarbieux, M. Montoya, M. Ribbrag, C. Mazur, C. Guet, P. Perrin, N.M. Maurel, J. Phys. Lett. 42 (1981) L437; M. Montoya, Thesis (Orsay, 1981).

11) C. Schmitt, A. Guessous, J.P. Bocquet, H.G. Clerc, R. Brissot, D. Engelhardt, H.R. Faust, F. Gonnenwein, M. Mutterer, H. Nifenecker, J. Pannicke, Ch. Ristori and J.P. Theobald, Preprint IKDA 84/2 (Darmstadt, 1984).

12) J. Trochon, G. Simon, J.W. Behrens and F. Brisard, Communication given at the International Conference on Nuclear Data for Basic and Applied Science, Santa Fe, New Mexico, USA, May 13-17, 1985.

13) G. Schutte, P. Möller, J.R. Nix and A.J. Sierk, Z. für Phys. A297 (1980) 289.

INFORMATION - THEORETICAL ANALYSIS OF WAVE PACKET DYNAMICS IN BOUND SYSTEMS

J. Brickmann[*]

The Fritz Haber Center for Molecular Dynamics, The Hebrew University
Jerusalem, Israel

* Permanent address: Institut für Physikalische Chemie
Technische Hochschule, Darmstadt, Petersenstr. 20, D-6100 Darmstadt, D.B.R.

1. Introduction

Understanding the basic mechanisms of intramolecular energy transfer is of fundamental importance in forming theoretical concepts of unimolecular reactions and other phenomena in molecular physics. There are many descriptions of unimolecular reactions, the dynamical extremes of which are the Slater and the Rice-Ramsperger-Kassel-Marcus (RRKM) models [1]. In all theories the intramolecular nuclear motion is treated as a set of oscillators (harmonic or non harmonic) which are coupled by linear or nonlinear coupling terms. A necessary (but not sufficient) condition for energy transfer between the vibrational modes in a classical picture is chaotic motion of the classical pseudoparticle. Statistical theories, however, can only be applied , if the classical trajectories cover the available phase space shell H = E (H: Hamiltonian, E: total energy) uniformly (ergodicity) within a time scale which is short with respect to other time scales of the molecular motion (like dissociations).

For molecular vibrations, the classical picture is, in general, not sufficient to describe the intramolecular vibrational energy transfer since quantum effects may play an important role. One possibility to link the classical mechanics to quantum mechanics is the study of quantum wave packets, $|\alpha\rangle$ which are initially located in phase as well as possible and study the dynamics of such states. In this paper it is demonstrated that one can obtain some general knowledge about the wave packet dynamics by applying information theoretic concept. Before considering these concepts for wave packets, we focus the discussion on the corresponding classical motion.

2. Probability distribution for classical motion

For classical dynamics the trend of a trajectory towards ergodicity (or a set of trajectories) can be measured, for example, by comparing ensemble averages $\langle f \rangle^e$ to time averages

$$\langle f \rangle_t = \lim_{T \to \infty} \frac{1}{T} \int_{t_0}^{t_0+T} f(s)dt \qquad (1)$$

where $f(s)$ is a dynamical variable of the system (a phase space function) and s is the phase coordinate denoting the aggregate at time t of generalized coordinates and momenta. It is convenient, to restrict the ensemble averaging to a partition of the phase space into nonintersecting sets Ω_i of a volume h^N - where N is the dimensionality of the position space - in order to compare the results to quantum motion. This partition leads to a probability distribution

$$W^e = \{W_i\} \qquad (2)$$

with individual probability W_i of finding the system in the i-th cell Ω_i of the phase space Γ

$$W_i = \int_{\Omega_i} \rho \, d\Gamma \quad . \qquad (3)$$

For a microcanonical ensemble the phase space density is simply a δ-function

$$\rho = \delta(H-E) \quad . \qquad (4)$$

For individual trajectories (or sets of trajectories) a distribution of time averaged probabilities $W^T = \{W_i^T\}$ can be determined in an analogous manner by averaging the times τ_i, the trajectories spend in the i-th cell [1]

$$W_i^t = \lim_{T \to \infty} (\tau_i/T) \qquad (5)$$

and the question occurs, how far is the distribution W^T away from the random distribution W^e, or how random is the motion of a trajectory (or of a set of trajectories).

3. Probability distribution for wave packets

The most natural counterpart for a trajectory (or a set of trajectories) in quantum mechanics is a wave packet, i.e. a non stationary quantum state which can be considered as the result of a physically reasonable preparation process. We restrict the considerations here to minimum uncertainty wave packets, Gaussians with the minimal spread in position and momentum space [2-4]. Such a wave packet can be considered as the result of a short time laser exitation from one electronic state to another via one or two photon processes in Condon approximation [1,2,6,7]. At t=0 the wave packet $|\alpha>$ is a well defined superposition of eigenstates $|\phi_j>$ of the Hamiltonian with coefficients c^{α}_i. The quantities $p^{\alpha}_i = |c^{\alpha}_i|^2$ are the probabilities to find the wave packet in the i-th eigenstate of the system. The distribution $P_{\alpha} = \{p^{\alpha}_i\}$ reflects on one hand the properties of the initial state preparation procedure and the properties of the eigenstates $|\phi_j>$ of the Hamiltonian $\hat{H}$ on the other. For a Hamiltonian with only regular eigenstates (i.e. in the case of a separable one, for example) the population probability distribution P_{α} will show a strong dispersion as a function of energy, i.e. there is only a relatively small number of contributing states in the Franck-Condon case [4,5,8,9]. If the $|\phi_i>$ are all irregular states the p^{α}_i will become a smooth and only slightly fluctuating distribution in energy space. Interpreting the eigenstates as a possible realization of the quantum space partitioning, the distribution P_{α} becomes a quantum analog of the time averaged distribution W^t (eq. 5) of a trajectory (or a set of trajectories) and one may ask - as in the classical case - for the most random (or statistical limit) distribution, which can be taken as a reference for a statistical behavior. Heller [8,9] has proposed such a conditional random distribution (CRD) $\{P^{\alpha T}_i\}$ wherein the individual probabilities $p^{\alpha T}_i$ are calculated with respect to the prior envelope or strength function $S_T(E)$ of the original distribution. $S_T(E)$ is the smoothed version of the fully strength function $S_{\infty}(E)$ with

$$S_{\infty}(E) = \sum_i p^{\alpha}_i \, \delta(E-E_i) \tag{6}$$

and can be written as

$$S_T(E) = \sum_i p_i^\alpha \, \Omega_T(E_i - E)$$

$$= \frac{1}{2\pi\hbar} \int_{-\infty}^{+\infty} \exp(iEt/\hbar) \, \hat{\Omega}_T(t) \, \mathrm{Tr}\,[\exp(-i\hat{H}t/\hbar) \, \rho^\alpha]dt \tag{7}$$

where $\hat{\Omega}_T(t)$ is a cutoff function which is equal to unity for $t < T$ and which decreases rapidly for $t > T$, ρ^α is the density operator corresponding to the wave packet $|\alpha\rangle$

$$\rho^\alpha = \sum_i |\phi_i\rangle \, p_i^\alpha \, \langle\phi_i| \tag{8}$$

and "Tr" indicates the trace. The CRD with Heller's approach becomes

$$p_i^{\alpha T} = S_T(E_i)/D_T(E_i) \tag{9}$$

where $D_T(E)$ is the density of states in the spectrum of $\hat{H}$ as can be calculated from the smoothed spectrum. For low density $D_T(E)$, the latter quantity is ill defined. Brickmann et al [5] have proposed a different procedure to generate a reference distribution $\{p_i^{\alpha R}\}$ which is based on the assumption that expansion coefficients C_i^α and C_{i+1}^α corresponding to energetically adjacent eigenstates $|\tilde{\phi}_i\rangle$ and $|\tilde{\phi}_{i+1}\rangle$, respectively, cannot drastically differ in amplitude for a Hamiltonian $\hat{H}$ with only irregular eigenstates. With irregular eigenstates we mean here such states, for which all physically reasonable properties are only smoothly varying with energy, but do not depend on other constants of motion. Such states are expected to correspond to Wigner distribution functions [10-12]

$$P_i^W(\underline{q},\underline{p}) = h^{-N} \int d\underline{x} \, \exp(-\underline{p}\cdot\underline{x}/\hbar) \, \tilde{\phi}_i^*(\underline{q}-\underline{x}/2) \, \tilde{\phi}_i(\underline{q}+\underline{x}/2) \tag{10}$$

(as quantum analogs to classical space densities) which are delocalized over the energetically available space region, i.e. these functions tend towards the microcanonical distribution in the classical limit [12]. The conditional random distribution $\{p_i^{\alpha R}\}$ of a given distribution can be generated from the p_i^α by a "probability diffusion" procedure [5] between neighbouring states.

It can be shown that such a procedure leads in any case to an increase of randomness. For n such diffusion steps (sharing of probabilities between n-states) with diffusion reflection at the lowest eigenvalue the CRD-components become [5]

$$p_j^{\alpha R} = n^{-1} (\frac{2}{\pi})^{1/2} \sum_i p_i^{\alpha} \exp \left[\frac{(j - \frac{1}{2})^2 + (i - \frac{1}{2})^2}{2n^2} \right] \cosh \left[\frac{(j - \frac{1}{2})(i - \frac{1}{2})}{n^2} \right] \cdot \quad (11)$$

These components agree with those of Heller eq. (9) in the high energy range if n and T are chosen conveniently but disagree in the low level density range. This disagreement is probably related to the ill defined density of states in the low energy range.

As in the classical case one has to ask how far is a given distribution away from the CRD or how far is a wave packet's motion away from a random motion ? To answer the question, one has to deal with criteria for the comparison of probability distributions.

4. How far is a wave packet's motion away from a random motion ?

The time evolution of an initial wave packet in a bound system with Hamiltonian is fully determined for all times if the expansion coefficients C_i^{α} with respect to the eigenstates of $\hat{H}$ and the spectrum $\{E_i\}$ of $\hat{H}$ are known. One simply has

$$|\alpha>_t = \exp (-i\hat{H}t/\hbar)|\alpha> = \sum_i \exp (-iE_i t/\hbar) \ C_i^{\alpha}|\phi_i> \quad (12)$$

i.e. $|\alpha>_t$ depends on both the coefficients and the spectral characteristics. Both properties, however, are not independent from each other for physically reasonable initial wave packets. Expanding a given minimumuncertainty wave packet with respect to a set of completely irregular eigenstates (with an irregular spectrum containing no low order resonances) will result in a set of coefficients C_i^{α} with all of the same order of magnitude while for an expansion with respect to regular eigenstates (with or without low order resonances in the spectrum) only a few coefficients may drastically differ from zero, i.e., the analysis of the distribution $p_i^{\alpha} = |C_i^{\alpha}|^2$ gives also qualitative information about the spectrum of $\hat{H}$. If $\{p_i^{\alpha}\}$ approaches the conditional random distribution, it is expected that the dynamics of the wave packet eq.(12) is related to an irregular spectrum. Individual wave packets with the same CRD but starting in different areas $\langle\alpha|\hat{q}\alpha\rangle$, $\langle\alpha|\hat{p}|\alpha\rangle$ of phase space then both spread in the same time scale over the available space and no recurrences are expected within physically interesting times. Two wave packets, starting in the regular region of the energy spectrum with a common CRD can move totally differently and, in general, do not spread over the available phase space region.

One can conclude, that statistical behavior of wave packets starting on different locations $\hat{q}$, $\hat{p}$ in phase space but with comparable energy $\hat{H}$ is expected, if the "information theoretical distance" between their probability distributions and corresponding CRD's are small and of comparable magnitude, while the time averaged properties corresponding to the wave packets motion become drastically dependent on the initial conditions if there are large fluctuations of these information theoretical distances. There is no unique way to measure the distance between two probability distributions. This point is discussed in a recent paper of Brickmann et al. [5]. The most frequently used method is related to the information entropy

$$S(p) = - \int p_i \ln p_i \quad .$$ (13)

We have applied these measures in a number of publications [4,5,10] to analyse the classical and corresponding wave packet's motion in two dimensional model systems for nonlinear oscillators with a Hamiltonian of the form

$$H = \frac{1}{2m} (p_x^2 + p_y^2) + \frac{m}{2} (\omega_1^2 x^2 + \omega_2^2 y^2) + \epsilon(x^2 y - \lambda y^3) \quad .$$ (14)

The probability distributions of two equienergetic wave packets, as well as the CRD's resulting from Heller's approach (eq.9) and the probability diffusion procedure (eq. 11) is shown in fig.1 and fig. 2. These wave packets start in an energy range where there is classical chaotic as well as regular motion. It is seen that the two distributions $\{p_i^\alpha\}$ are quite different while the CRD's are nearly identical. In the first packet (fig. 1) the information entropy difference

$$\Delta S = S(p^{\alpha R}) - S(p^\alpha)$$ (15)

is much larger than in the second case. For wave packets, starting in an
energy range with almost complete chaotic motion, all ΔS values approach a
common small value [5].

There is no one-to-one correspondence between the information entropy
calculated from the time averaged probabilities $\{W_i^t\}$, eq. (5), for a trajectory
starting at the phase space coordinates (p,q) to the entropy of a minimum
uncertainty wave packet $|\alpha\rangle$ with location $\langle\alpha|\hat{p}|\alpha\rangle$, $\langle\alpha|\hat{q}|\alpha\rangle$ identical to the
classical coordinates, but such a correspondence is not to be expected.
We are presently checking, under what conditions such a correspondence can
be established when a wave packet's motion is compared to a swarm of trajectories with a distribution in phase space corresponding to the wave packets
distribution. However, a nice correspondence between the ergodic limit
entropy $S(W^e)$ and the entropy $S(p^{\alpha R})$ of the CRS's for wave packets located
initially in the center of the phase space cells Ω_i was found

$$S(p^{\alpha R}) = S(W) + \delta S \tag{16}$$

where δS is a small fluctuation relative to the absolute value of S.
This result indicates, that the probability distribution $\{p_i^{\alpha R}\}$ is indeed a
quantum analog to the classical ergodic limit distribution $\{W_i\}$.

Acknowledgement

I like to thank Prof. E. Heller for stimulating discussions and
Dr. P. Cribb for carefully reading the manuscript. This work was supported
by the Deutsche Forschungsgemeinschaft, Bonn and the Fonds der Chemischen
Industrie, Frankfurt. The Fritz Haber Research Center and my stay in Israel
is supported by the Minerva Gesellschaft für die Forschung in BH, München.
I wish to acknowledge the kind hospitality shown to me during my stay in the
Fritz Haber center.

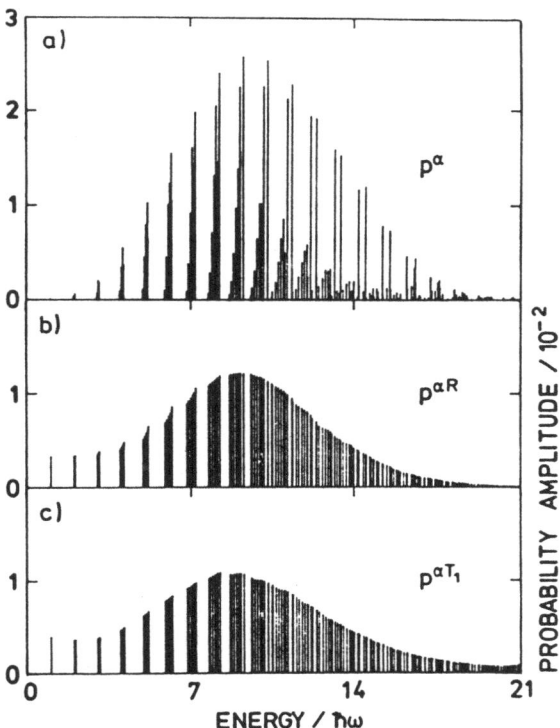

Figure 1: Probability distributions p^α, $p^{\alpha R}$(eq. 11) and $p^{\alpha T}$ (eq. 9) for a minimum uncertainty wave packet $|\alpha\rangle$ with $\langle\alpha|q|\alpha\rangle = (0.25, -2.39)$, $\langle\alpha|p|\alpha\rangle = (-2.25, 0.22)$ with respect to a Hamiltonian

$$H = \frac{1}{2}(p_x^2 + p_y^2 + x^2 + 1.1y^2) + 0.1089\ (x^2 y - 10y^3/33)$$

($\hbar = 1$, $\omega = 1$, $m = 1$ units)

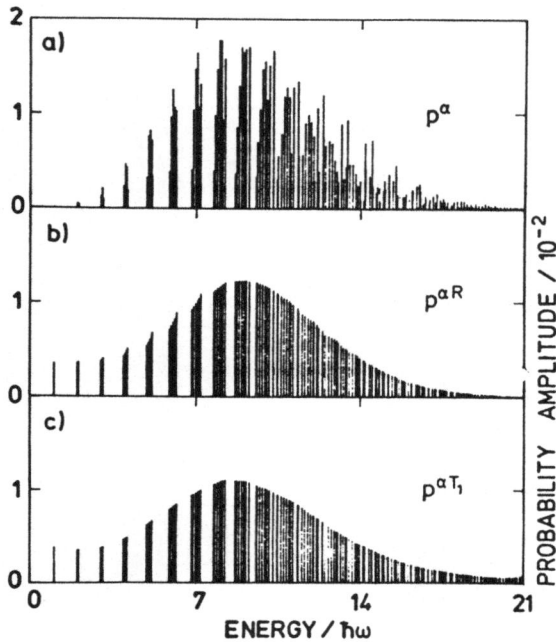

Figure 2: Probability distributions p^{α}, $p^{\alpha R}$ and $p^{\alpha T}$ as in fig. 1 for

$$\langle\alpha|q|\alpha\rangle = (0.25, -2.34), \qquad \langle\alpha|p|\alpha\rangle = (2.75, -2.16)$$

References

1) J. Brickmann, R. Pfeiffer, P.C. Schmidt, Ber. Bunsenges. Phys. Chem. <u>88</u> (1984) 382-397 and references cited therein.

2) J. Brickmann, P. Russegger, J. Chem. Phys. <u>75</u> (1981) 5744

3) J. Brickmann, J. Chem. Phys. <u>78</u> (1983) 1824

4) J. Brickmann, P.C. Schmidt, Inter. J. Quantum Chem. <u>23</u> (1983) 47

5) J. Brickmann, P. Hofmann, P.C. Schmidt, "Measures of Stochasticity for Non-Stationary Initial Preparations of Coupled Oscillators", Intern. J. Quantum Chem. 1985, in press

6) E.J. Heller, J. Chem. Phys. <u>68</u> (1978) 2066, J. Chem. Phys. <u>68</u> (1978) 3891

7) P. Cribb, J. Brickmann, to be published

8) E.J. Heller, J. Chem. Phys. <u>72</u> (1980) 1337

9) E.J. Heller, MJ. Davies, J. Chem. Phys. <u>86</u> (1982) 2118

10) J. Brickmann, P.C. Schmidt, Chem. Phys. Lett. <u>104</u> (1984) 591

11) M. Tabor, Adv. Chem. Phys. <u>46</u> (1981) 73

12) M.V. Berry, J. Phys. <u>A10</u> (1977) 2083

THE GAUGE INVARIANT METHOD FOR QUANTIZATION AND TUNNELING IN HAMILTON SYSTEMS

E. Caurier

Centre de Recherches Nucléaires, B.P. 20, 67037 Strasbourg Cedex, France

M. Ploszajczak

Niels Bohr Institute, DK-2100 Copenhagen Ø, Denmark

Institute of Nuclear Physics, 31-342 Krakow, Poland

S. Drozdz

Institute of Nuclear Physics, 31-342 Krakow, Poland

An application of the Hamiltonian dynamics for bounded quantal systems requires the formulation of appropriate boundary conditions which is provided, for instance, by the regularity and single-valuedness (RSV) principle for gauge invariant solutions of the Schrödinger time-dependent variational principle [1,2]. This method can be applied in any manifold of the Hilbert space [3], both for a bound state problem [1,4] as well as for a tunneling [5]. It provides a variational approach to the quantum dynamics and as such, it depends mainly on the chosen variational manifold $\{\Psi(\vec{q},\vec{p})\}$ for canonical conjugate, classical variables $\vec{q}$ and $\vec{p}$. RSV method underlines particularly the role of $\vec{q},\vec{p}$ as labels of the evolving wave-function and the interpretation of the expectation value $\mathcal{H}(\vec{q},\vec{p}) = \langle\Psi(\vec{q},\vec{p})|\hat{H}|\Psi(\vec{q},\vec{p})\rangle$ of the quantal many-body Hamiltonian as the classical Hamiltonian [6]. Consequently, the Schrödiger time-dependent variational principle yields quite naturally Hamilton eqs. for the time-develement of $\vec{q},\vec{p}$ and hence $\Psi(\vec{q},\vec{p})$. Increasing the number of labels q,p is equivalent to a gradual changing from a classical to a quantum theory. Finally, in the full Hilbert space, one arrives at the time dependent Schrödinger eigenstates and only them [1,4]. (The details of the RSV quantization method can be found in our earlier works [1,5,7]. The relation between labels $\vec{q},\vec{p}$ and phase-determined gauge-invariant wave-function $\Psi_G(x,\vec{q},\vec{p})$ is regular and single-valued if for any closed trajectory C_σ in the label space:

$$\oint_{C_\sigma} \vec{p}d\vec{q} = 2\pi n_\sigma \hbar \qquad n_\sigma = 0,1,\ldots \qquad (1)$$

This property is assumed to hold for any physical bound state and can be
rewritten in the form of a quantization condition

$$I_i = (2\pi)^{-1} \int_{C_i} \vec{p}d\vec{q} = n_i \hbar \qquad \begin{array}{l} n_i = 0,1,\ldots \\ i = 1,\ldots,n \end{array} \qquad (2)$$

for components of the action integral $\vec{I} = [I_1,\ldots,I_n]$ where $2n$ is the dimensionality of the label space. This however is possible if one find n topolo-
gically independent basic closed curves $\{C_i\}$: $C_\sigma \equiv \sum_{i=1}^{n} M_i C_i$ for $[M_1,\ldots,M_n]$
being a set of integer numbers. One should notice that in contrast to quasi-
classical methods [8], the quantization path in RSV quantization method is
constructed in the space of phase determined wavefunction and not in the
classical configuration space. Consequently, the Maslov index [9] does not
enter into expressions (1), (2) and the lowest quantized solutions $\vec{I}^{(i)} =$
$[I_1 = 0,\ldots, I_i = n_i\hbar,\ldots,I_n = 0]$ $(i=1,\ldots,n)$ should be time-periodic.
Those "stationary" solutions are particularly interesting since they offer
a generalization of RPA modes for infinitesimal oscillations. In a following
discussion we present a simple approach for selecting in non-separable
Hamilton systems the periodic trajectories which represent the topologically
independent closed circuits on the invariant torus. Obviously, we anticipate
an existence of invariant tori at least in the part of a label space $\{\vec{q},\vec{p}\}$.
For each excitation energy E^* the collective motion is initiated on the
equienergy surface $\mathcal{H}(\vec{q},\vec{p} = 0) = \langle\Psi(\vec{q},\vec{p} = 0)|\hat{H}|\Psi(\vec{q},\vec{p} = 0) = E^*$ where $\Psi(\vec{q},\vec{p})$
is the variational wavefunction and $\hat{H}$ is a quantum many-body Hamiltonian.
The numerical procedure of finding the initial points goes as follows.
We begin at low excitation energy $E^*_{(1)}$ by selecting the reference initial
point on $(\vec{q},\vec{p} = 0)$ which for example can be found by determining RPA modes.
Position of each point on $\mathcal{H}(\vec{q},\vec{p} = 0) = E^*_{(1)}$ can be conveniently assigned by
$N-1$ polar angles $\theta_1,\ldots,\theta_{N-1}$. Then keeping those angles unchanged the
excitation energy is increased $E^*_{(1)} \to E^*_{(2)} = E^*_{(1)} + \Delta E$ and we determine
new initial points on $\mathcal{H}(\vec{q},\vec{p} = 0) = E^*_{(2)}$. In general, those points may not
give rise to the least coupled modes and one should try to find the better
points. For that, in the neighbourhood of each reference point we initiate
the evolution we calculate the difference $\Delta_{i,f} = |\vec{q}_{(i)} - \vec{q}_{(f)}|$ of initial and
final $\vec{q}$ - values at the first minimum of this quantity, i.e. after one
oscillation. Obviously if $\Delta_{i,f} = 0$ then $\vec{p}_{(i)} = \vec{p}_{(f)} = 0$ and we have found a
decoupled variable. However, in general $\vec{p}_{(i)} \neq \vec{p}_{(f)}$ and the final point
$\vec{q}_{(f)}$ is situated at the nearby caustic line for which one of the components
of the momentum $\vec{p}$ is equal zero. Minimizing function $\Delta_{i,f}$ around each refe-

rence point separately, one can find easily an optimal set of angles $\theta_1,\ldots,\theta_{n-1}$ for each mode. For those optimized modes j (=1,...,n) we now calculate action integrals

$$I^{(j)} = (2\pi)^{-1} \int_{\vec{q}(i)}^{\vec{q}(f)} \vec{p}d\vec{q} \qquad (3)$$

and check whether they satisfy the RSV quantization condition. For those which do not obey the RSV condition, we change the excitation energy and the newly found optimal angles provide us with the necessary points of reference. We continue with this numerical procedure until quantized energy of each decoupled or least-coupled mode is found. An identification of "collective" variables by comparison with those at ifinitesimal excitation is straightforward provided the variation op optimal initial points with energy is continuous. Typical examples of periodic trajectories as obtained in the above method, are shown in fig.1. The upper picture shows coupled x,z shape oscillations of ^{8}Be which are described using TDHF approach for a deformed oscillator (DO) model and the B1-force [10]. Calculations are performed at $E^* = 10,20,30$ MeV and the highest energy modes touch the energy surface $\mathcal{H}(\vec{q},\vec{p} = 0)$. (Details of this mode can be found in ref. [7]). A small cross in the middle of a picture denotes the ground state deformation of ^{8}Be. Arms of this cross point to the directions of RPA modes. Below, similar results are shown for the TDHF evolution of 60 particles in the SU(3) model. (Details of the TDHF + SU(3) model are give in ref. [11]. Same as before the results at $E^* = 0.2$, 0.3, 0.4 are plotted in the RPA frame of reference. (In both cases the non-separable Hamiltonian system is evolved in 2N=4 variables.) Periodic trajectories in fig. 1 are displayed with respect to both the equilibrium configuration and the RPA mode. Energies of few lowest quantized states in TDHF + DO and TDHF + SU(3) models are compared in Table 1 with the corresponding RPA solutions for infinitesimal oscillations.

(n_1, n_2)	^8Be TDHF +RSVQ	RPA	^{12}C TDHF +RSVQ	RPA	^{20}Ne TDHF + RSVQ	RPA	^{28}Si TDHF + RSVQ	RPA	N-60 χ^{-10} TDHF + RSVQ	RPA	N-60 χ^{-100} TDHF + RSVQ	RPA
(0,1)	14.57	17.67	27.68	34.56	16.27	16.98	28.03	29.69	0.1596	0.1715	1.7543	1.9075
(1,0)	26.37	33.01	17.6	20.59	24.76	26.13	16.93	17.735	0.1968	0.2060	1.8069	1.9408
(0.2)	26.18	35.34	44.69	69.12	31.17	33.96	52.64	59.38	0.2992	0.3430	3.3402	3.8150
(2,0)	-	70.68	30.27	41.18	46.87	52.26	32.36	35.47	0.3851	0.4120	3.3981	3.8816

Table 1

It is interesting to notice significant energy differences between quantized
levels in the two methods which reflect qualitative differences in fig.1
between DO or SU(3) periodic trajectories and the respective RPA modes.
The large decrease of energy due to anharmonicities is often found for a
higher mode (see e.g. ^8Be, ^{12}C). The method of finding periodic trajectories
can be easily extended to $(n > 2)$ higher dimensions.
The above results are also relevant for a qualitative discussion of the mode
coupling in a quantum tunneling process. A motion inside a barrier can be
described by the Hamilton eqs. in the imaginary time or, equivalently, in the
inverted potential. As a result of the mode-mode coupling the collective
variable associated with a tunneling is strongly modified and the tunneling
width is damped even though the barrier height is unchanged. Detailed
investigations of those and related topics in the context of a quantum
dissipation provides a fascinating challenge for the future studies in both
the nuclear and molecular physics.

References

1) E. Caurier, S. Drozdz and M. Ploszajczak, Phys. Lett. 134B (1984) 1

2) K.K. Kan, J.J. Griffin, P.C. Lichtner and M. Dworzecka,
   Nucl.Phys. A332 (1979) 109

3) K.K. Kan, Phys. Rev. C24 (1981) 279

4) S. Drozdz, J. Okolowicz and M. Ploszajczak, Phys.Let. 115B (1982) 161

5) E. Caurier, S. Drozdz and Ploszajczak, Phys. Lett. 150B (1985) 1
   and J. de Phys. C6 (1984) 361

6) J.R. Klauder, J. Math. Phys. 8 (1967) 2392

7) E. Caurier, S. Drozdz and M. Ploszajczak, Nucl. Phys. A425 (1984) 233

8) A. Einstein, Verth. Dt. Phys.? Ges. 19 (1917) 82;
   M.L. Brillouin, J. de Phys. 7 (1926) 353;
   J.B. Keller, Ann. of Phys. 4 (1958) 180

9) V. Maslov, Théorie des perturbations (Dunod, Paris, 1972)

10) B.M. Brink and E. Boeker, Nucl.Phys. <u>A91</u> (1967) 1.

11) R.D. Williams and S.E. Koonin, Nucl. Phys. <u>A391</u> (1982) 72

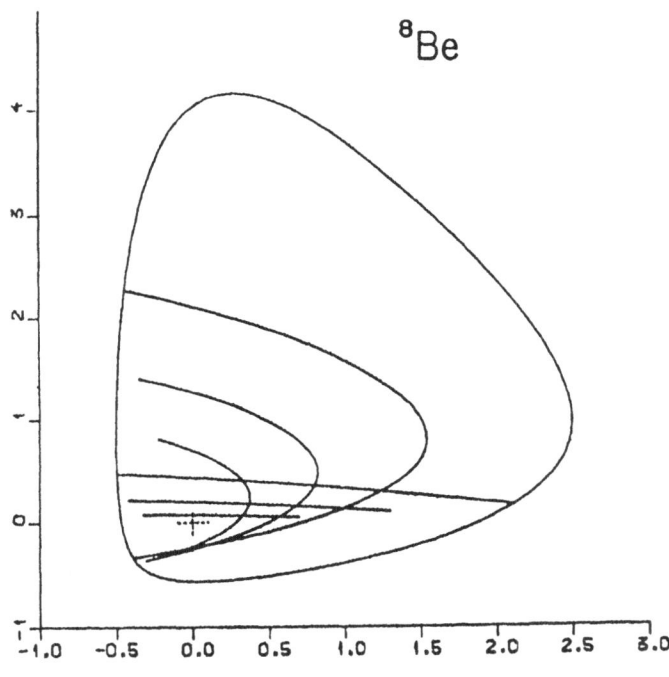

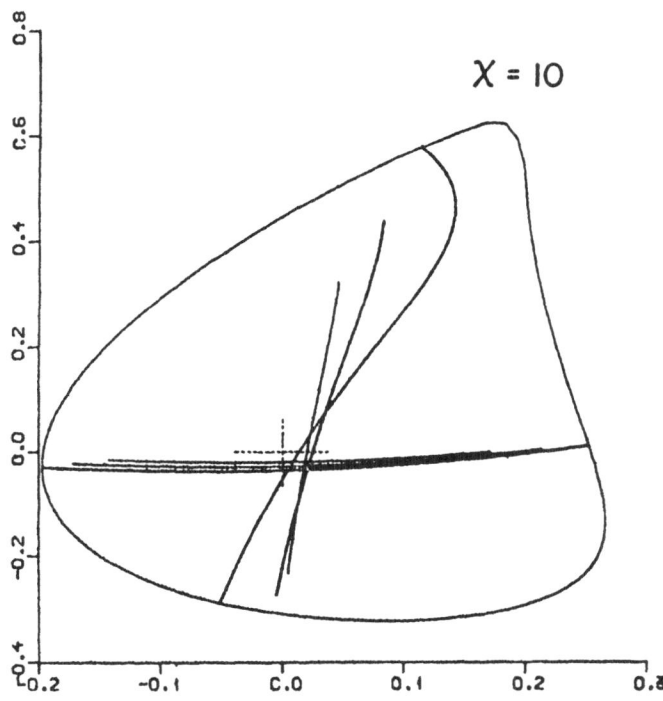

PARTICLE-SOLID INTERACTIONS

C. Cerjan

Lawrence Livermore National Laboratory, University of California,
POB 808,L-487, Livermore, CA 94550, U.S.A.

The solution of the exact time dependent Schrödinger equation for particle-surface scattering provides a direct and natural means for studying the dynamics of this interaction. Atomic or molecular scattering at thermal energies from crystalline surfaces presents several interesting problems which are of both practical and fundamental interest. The scattering experiment might provide information about the structure of the surface and its vibrational characteristics that would be difficult to obtain by other experimental techniques. The transfer of energy by a flowing gas to a surface is important in gas flow cooling since the boundary conditions for the temperature are established at the surface. Knowledge of the temperature jump at the surface is needed as input for the continuum models used to evaluate cooling rates. Also, the rate of desorption of various species from a crystalline surface as a function of temperature is often a quantity that must be known in many experimental situations, yet it remains quite difficult to approximate.

Time dependent methods offer a natural setting for the study of temperature and vibrational effects at surfaces. The description of the surface motion is much more readily expressed and understood as an explicitly time dependent potential. Since the wave-like nature of the projectile is often important, it is necessary to solve the Schrödinger equation for the translational motion of the projectile, which is then coupled to either a classical or quantal "bath" of phonons, depending upon the physical situation. Traditional closed-coupled channel methods are useful for static considerations, yet they are often quite difficult to incorporate into a multiple phonon picture. So, as a first step in the study of these interactions, the time dependent Schrödinger equation is used for the propagation; coupling of the particle to a bath is described either classically or quantum mechanically. Furthermore, this approach is closely allied to the usual classical descriptions of particle propagation and these insights may be used in the full quantum mechanical treatment or used for comparison purposes.

The model thus falls into two parts: the solution of the Schrödinger equation and the introduction of the appropriate time dependent couplings. To solve the Schrödinger equation, a Fast Fourier Transform (FFT) technique is used. Specifically, the kinetic energy operator is calculated by first transforming the wave function to momentum space, mutiplying the result by k^2 and then synthesizing the product. In this technique, the potential becomes a multiplication operator. The time propagation is handled by a finite differencing scheme or more sophisticated techniques.

The advantages of this technique are that it has less numerical dispersion than simple finite differencing algorithms for the kinetic energy operator; it is easy to program, much like classical or semiclassical techniques; and it is highly adaptable to parallel processing computers. The disadvantages are that it might require large mesh sizes which renders some problems intractable, and, to some extent, it is restricted to Cartesian coordinates. Also, it might not be as accurate as closed-channel methods. It should also be noted that many computers have machine language or optimized FFT routines available. In some instances these routines can improve computing times by a factor of two or more.

The remaining part of the problem - specification of the time dependent potential - can be approached either classically or quantum mechanically. The only constraint is that the static potential contain realistic features. The potential must be exponentially repulsive at small separations and should have polarization and corrugation effects if an accurate comparison is desired. The time dependence of the surface and its coupling to the translating particle may be introduced by self-consistently coupling a Langevin description of the surface and bulk for a classical solid. A semi-group approach for the construction of a Liouville dissipative operator may be used for the quantal solid.

The parameters needed for the solid and its surface must be introduced empirically. It is known, though, that the results are not very sensitive to the choice of these parameters. The coupling of the particle's motion to the surface is more arbitrary and can lead to qualitatively different effects. Results for one dimensional desorption and three dimensional scattering from a dynamic classical surface reveal the utility of the method. It should be noted that an absorbing boundary condition was used to attenuate the wavepacket at the edge of the grid thus permitting long sampling times.

As a further application, diatom scattering from a corrugated static surface was examined to test an interaction representation for the FFT method. That is, the idea is to perform a mixed time dependent, closed-coupled channel evaluation to reduce the dimensionality of the full problem. Scattering of molecular hydrogen and HD were examined. Additionally, a new time evolution operator (Chebyshev expansion) was introduced which provides high accuracy.

Finally, the thermal accomodation of helium provides an example of the use of a quantum mechanical description of the surface. The specific problem addressed is the evaluation of the equilibrium accomodation coefficient of helium as a function of temperature. The accomodation coefficient is a dimensionless measure of the amount of energy tranferred to the surface. Current theories emphasize time independent approaches and are seriously deficient at low temperatures. To study this problem, a semi-group method was used to construct the appropriate dissipative terms. One and two phonon processes coupled to a harmonic bath were introduced and the dissipation within the bath is then uniquely defined. The bath is initially at equilibrium with a boson number distribution and remains at equilibrium until the scattering particle begins to interact strongly with it.

In order to mimic the experimental situation, a Maxwell-Boltzmann distribution is chosen for the initial wavepacket. As the temperature is lowered, the probability that the particle remains in the well grows -- in contrast to the similar time independent perturbative approaches. This qualitative feature alone displays the utility of time dependent methods.

Much can be learned from the explicitly time dependent techniques. At present, it is still difficult to analyse some features of the scattering process (such as the partial differential cross-section or average energy transfer) but these preliminary results indicate that the approach deserves further consideration, especially for particle-surface interaction phenomena.

TIME-DEPENDENT DENSITY FUNCTIONAL THEORY

R.M. Dreizler and H. Köhl

Institut für Theoretische Physik, Johann Wolfgang Goethe Universität
Robert-Mayer Straße 8-10, Postfach 11 19 32, 6000 Frankfurt/Main 11, D.B.R.

Suppose we want to discuss an explicitly time dependent many fermion system, characterized by a Hamiltonian of the form

$$\hat{H}(t) = \hat{T}+\hat{V}+\hat{W} = \sum_i (\hat{T}_i+\hat{V}_i(t)) + \frac{1}{2}\sum_{i\neq k} \hat{W}_{ik}$$

and an initial state $|\phi(t_0)\rangle$. We ask the question, whether it is possible to provide an exact description of this system in terms of the one particle density $\rho(\underline{r},t)$ alone? The question is to be answered on two levels, one being the level of an answer in principle (the level of existence theorems) the second being the practical level (if the answer on the first level is positive). Concerning the first level the following remarks can be offered. Runge and Gross [1] were able to prove the following theorem:
There exists a three component density functional $\vec{P}[\rho]$ such that the exact one particle density can be calculated as a solution of the equations

$$\partial_t \rho(\underline{r},t) + \text{div } \vec{j}(\underline{r},t) = 0$$

$$\partial_t \vec{j} = \vec{P}[\rho]$$

with appropriate initial conditions $\rho(\underline{r},t_0)=\rho(\underline{r})$ and $\vec{j}(\underline{r},t_0) = \vec{j}(\underline{r})$.
The assumption required for the proof of the theorem is: The one particle potential function $V(\underline{r},t)$ can be expanded in a Taylor series around the initial point t_0, i.e. this function has a smooth analytical structure in time.

The theorem implies that the time dependent many particle problem can be transcribed (in principle) into the framework of hydrodynamics. As another option [1] one can set up, loosely speaking due to the fact that any one particle density or current density can be expanded in terms of suitable effective orbitals e.g.

$$\rho(\underline{r},t) = \sum_{n=1}^{N} \phi_n^*(\underline{r},t)\phi_n(\underline{r},t),$$

a set of effective single particle wave equations for these orbitals (a time dependent Kohn-Sham scheme [2]) which (in principle) go beyond the TDHF limit.

The theorem can be viewed as an extension of the Hohenberg-Kohn theorem [3] of stationary density functional theory, where the existence of a universal functional $E_0[\rho]$ for the ground state energy is demonstrated. The connection is, however, nontrivial as the stationary case is based on the Rayleigh-Ritz principle, while in the case of TD systems the discussion is based on the action principle. The action integral

$$A = \int_{t_0}^{t} <\Phi(t)|i\hbar\partial_t - H|\Phi(t)>dt$$

provides only stationary points rather than extremal points. Nonetheless the stationary result can be shown to result from the time dependent one in the appropriate limit.

This discussion of an existence theorem does not offer any help for practical issues. Attempts to realize the principle have only started. In order to approach the practical side, we go back to the simpler, but nonetheless interesting problem of a system of noninteracting fermions moving in a common time dependent potential.

$$t \leq 0: \hat{H}(t\leq 0) = \hat{T} + \hat{V}(t\leq 0) = \sum_i (\hat{T}_i + \hat{V}_i(t=0)) = \sum_i \hat{h}_i(t\leq 0)$$

$$t > 0: \hat{H}(t) = \hat{T} + \hat{V}(t) = \sum_i (\hat{T}_i + \hat{V}_i(t)) = \sum_i \hat{h}_i(t) \quad .$$

To be specific we assume that for $t \leq 0$ the system is in its groundstate. The one particle density operator up to time $t = 0$ is then

$$\hat{\rho}(t\leq 0) = \Theta(E_F - \hat{h}(t\leq 0)) \quad .$$

The Fermi energy E_F is determined by specifying the particle number. The density operator for $t > 0$ is <u>not</u> obtained by replacing $\hat{V}(t=0)$ by $\hat{V}(t)$ but rather by proper time evolution

$$\hat{\rho}(t) = \hat{U}(t)\ \hat{\rho}(t=0)\ \hat{U}^+(t).$$

Introducing the operator

$$\hat{I}(t) = \hat{U}(t)\ \hat{h}(t=0)\ \hat{U}^+(t)$$

one can then write

$$\hat{\rho}(t) = \Theta(E_F - \hat{I}(t))\quad .$$

Any further discussion of the time dependent density and the current density defined as

$$\hat{\vec{j}} = \frac{1}{m}\ (\hat{\vec{p}}\hat{\rho} + \hat{\rho}\hat{\vec{p}})$$

by hydrodynamical schemes ought to be based on the operator $\hat{I}(t)$. This operator can alternatively be characterized by the equation

$$i\hbar\partial_t \hat{I}(t) + [\hat{I}(t),\hat{h}(t)] = 0$$

$$\hat{I}(t=0) = \hat{h}(t=0)$$

and may be recognized as the invariant operator first introduced by Lewis and Riesenfeld [4]. These authors also demonstrated the fact that $\hat{I}(t)$ is an isospectral deformation of $\hat{h}(t=0)$

$$\hat{I}(t)|n,t> = \lambda_n|n,t>\quad ,\quad \partial_t\lambda_n = 0$$

and in view of the initial condition

$$\lambda_N + E_n \quad .$$

It is exactly this property which makes the operator $\hat{I}(t)$ useful for the discussion of the density operator.

Unfortunately, fairly limited explicit knowledge of such invariants is available so far. Exact invariants are known for a rather small class of problems, including the TD harmonic oscillator. It is also known that there exists only a specific class of potentials, for which the invariant is a quadratic form of the momentum operator [5,13]. In general a more involved momentum dependence is to be expected [10]. For the situation that the Hamiltonian can be expanded in terms of the operators of a closed Lie algebra

$$\hat{h}(t) = \sum_n h_n(t)\hat{\Gamma}_n \quad \text{with} \quad [\hat{\Gamma}_n, \hat{\Gamma}_m] = \sum_\ell \lambda^\ell_{nm}\hat{\Gamma}_\ell,$$

the invariant can be expressed by a corresponding expansion

$$\hat{I}(t) = \sum_n a_n(t)\hat{\Gamma}_n$$

where the coefficients a_n are determined by a set of first order differential equations [6,10].

Some use can be made of this limited information. One can for instance demonstrate [7], using second order gradient expansion techniques, that the known invariants quadratic in $\hat{p}$ lead to a set of hydrodynamical equations consisting of

$$\partial_t \rho + \text{div}(\rho \vec{\nabla}\chi) = 0$$

$$\partial_t \chi + \frac{1}{2m}(\vec{\nabla}\chi)^2 + \frac{\delta}{\delta\rho} E_{stat}[\rho] = 0$$

suggested as early as 1933 by Bloch [8]. $E_{stat}[\rho]$ is the (second order) density functional of the stationary theory with the replacement $\rho \rightarrow \rho(t)$. The fact that the scheme involves irrotational flow can be traced back to the special $\hat{p}$-dependence of the invariant used. Numerical solution of hydrydynamical equations is not very simple, but we were able to provide some pilot calculations on the basis of the Bloch scheme and variants covering photoabsorption in atoms and ionatom collisions at higher energies [9].

In addition one can demonstrate that the kinetic energy density $\tau(\rho)$ of the time dependent system separates into an intrinsic and into a collective part

$$\tau(\rho) = \tau_{intr.}(\rho) + \frac{\vec{j}^2}{2m\rho}$$

for times t immediately after the time-dependence has been switched on [16].

In view of this situation one should look also for alternatives. We just point out two of them.

A - The semiclassical scheme ($\hbar=0$)

Considering the Weyl transform of the invariant, which satisfies the equations

$$\partial_t I(p,q,t) + \frac{2}{\hbar} \sin \frac{\hbar}{2} (\vec{\nabla}_p^h \vec{\nabla}_q^I - \vec{\nabla}_p^h \vec{\nabla}_q^I) I . h = 0$$

$$I(p,q,t=0) = h(p,q,t=0),$$

one readily finds for the classical limit ($\hbar=0$)

$$\partial_t I_{c\ell} + \vec{\nabla}_p h . \vec{\nabla}_q I_{c\ell} - \vec{\nabla}_q h . \nabla_p I_{c\ell} = 0$$

$$I_{c\ell}(t=0 = h(t=0).$$

The Weyl transform of the density operator (in the $\hbar=0$-limit) is the Wigner distribution function

$$f_{c\ell}(p,q,t) = 2\Theta(E_F - I_{c\ell}(p,q,t))$$

where the factor 2 takes care of spin degeneracy. On the basis of these remarks one can envisage the following programme

a) Determine the time propagation of the hypersurface

$$E_F = I_{c\ell}(\underline{p}, \underline{q}, t)$$

in phase space accordingly to the classical differential equation by numerical methods.

b) Calculate the time development of the density

$$\rho_{c\ell}(\underline{q}, t) = \frac{1}{h^3} \int d^3 p \, f_{c\ell}(\underline{p}, \underline{q}, t)$$

and other expectation values, as for instance $E_{kin}(t)$ and $E_{pot}(t)$.

We have applied this scheme to a number of model situations [11] and (using Monte Carlo techniques) to the discussion of cross sections for the production of recoil charge states in energetic ion-atom collisions [12].

<u>B</u> - The concept of local invariants

The starting point in this case is an exact formula for the density matrix of a systems of noninteracting fermions evolving from the ground state in an external time dependent field [13]

$$\rho_t(q|q'; E_F) = \frac{1}{2\pi i} \int_{c-i\infty}^{c+i\infty} d\zeta \, \frac{e^{E_F \zeta}}{\zeta} \oint_{q'}^{q} e^{\frac{i}{\hbar} \int_0^{-i\hbar\zeta} [p \cdot \dot{q} - I(p,q,t)d\tau]} \, DpDq$$

$I(p,q,t)$ is the Weyl transformation of the operator $\hat{I}$, as defined above. A phase space path integral is necessary due to the complicated p-dependence of $I(p,q,t)$.

Path integrals are notoriously difficult to handle in practical situations, but one could attempt to achieve some success with the following approximation, which is reasonable at least for short time propagation and some other special situations. We expand the potential about a given arbitrary point $\bar{q}$ up to second order, leading to

$$h(p,q,t) = \frac{p^2}{2m} + V(\bar{q}, t) + V'(\bar{q}, t)(q-\bar{q}) + \frac{1}{2} V''(\bar{q}, t)(q-\bar{q})^2 \quad .$$

For this Hamiltonian the invariant can be calculated exactly and the propagator can be evaluated by saddle point methods. Some model calculations along these lines can be found in ref. [14]. We mention that several other results can be derived from the path integral formula above.

We note [14] that E. Hellers wave packet method [15] can be derived by quite similar means. The eigenstates of the eigenvalue problem defined by the above <u>local</u> time dependent invariant are exactly the basis states of Coalson and Heller [17].

References

1) E. Runge and E. Gross, Phys. Rev. Lett. <u>52</u> (1984) 997

2) W. Kohn and L.J. Sham, Phys. Rev. <u>A140</u> (1965) 1133

3) P. Hohenberg and W. Kohn, Phys. Rev. <u>B136</u> (1964) 864

4) H.R. Lewis and W.B. Riesenfeld, J. Math. Phys. <u>10</u> (1969) 1458

5) H.R. Lewis and P.G. Leach, J. Math. Phys. <u>23</u> (1982) 2371

6) H.J. Korsch, Phys. Lett. <u>74A</u> (1979) 294

7) A. Henne and R.M. Dreizler, to be published

8) F. Bloch, Z. Phys. <u>81</u> (1933) 363

9) P. Malzacher and R.M. Dreizler, Z. Phys. <u>A307</u> (1982) 211
   M. Horbatsch and R.M. Dreizler, Z. Phys. <u>A300</u> (1981) 119
   M. Horbatsch and R.M. Dreizler, Z. Phys. <u>A308</u> (1982) 392

10) H. Kohl and R.M. Dreizler, submitted to Phys. Lett.

11) M. Horbatsch, H. Kohl and R.M. Dreizler, submitted to Z. Phys.

12) M. Horbatsch and R.M. Dreizler, submitted to Phys. Rev. Lett.

13) H. Kohl and R.M. Dreizler, Phys. Lett. <u>98A</u> (1983) 95

14) H. Kohl and R.M. Dreizler, Journ. de Physique <u>C6</u> (1984) 35

15) E. Heller, J. Chem. Phys. <u>62</u> (1975) 1544

16) H. Kohl and R.M. Dreizler, in preparation

17) See the contributions to this volume.

TIME-INDEPENDENT WAVE PACKET
THEORY OF COLLISIONS

B.G. Giraud

Service de Physique Théorique, C.E.N. Saclay, 91191 Gif-sur-Yvette, France

1. Introduction

The results briefly described in this lecture are the product of a many year collaboration with M.A. Nagarajan, Y. Abe, P. Amiot, C. Noble and I.J. Thompson. Three main aspects can be listed, namely
i) a wave packet representation of channels and reaction mechanisms,
ii) a variational approach to the calculation of transition amplitudes and
iii) a mean field approximation. The theory has now reached a stage where practical collision amplitudes become available.

2. The wave-packet representation

The physical system under study is an N-particle system with coordinates r_i, momenta p_i and masses m_i, i=1...N respectively. The wave packets introduced in this theory are specifically designed to accomodate many-body calculations: they have to be products of single particle wave packets. In momentum representation, for instance, we consider:

$$\chi(p_1 \ldots p_N) = \mathcal{H} \chi_1(p_1) \ldots \chi_N(p_N), \tag{2.1}$$

where the χ_i's can be Gaussians for instance. The operator $\mathcal{H}$ is a product of symmetrizors and/or antisymmetrizors when bosons and/or fermions are considered. It will be omitted for simplicity in the following.

The factorization of many-body integrals provided by the representation {χ} is an enormous advantage in practical calculations. It is also known that the set {χ} makes an (overcomplete) set in the Hilbert space. These two advantages may seem contradictory, however, with the fact that the physics under study is usually not related to single particle properties. Most often the physics focusses on one (or several) observables $R_\alpha(r_1 \ldots r_N, p_1 \ldots p_N)$ and/or their conjugate observables $P_\alpha(r_1 \ldots r_N, p_1 \ldots p_N)$ and the temptation is great to set up a formalism in terms of R_α, P_α and complementary variables ξ, π rather than the initial variables $\{r_i, p_i\}$

There is in fact no need to face the unwieldy transformation $\{r_i, p_i\} \leftrightarrow \{R_\alpha, P_\alpha, \xi, \pi\}$ and the technical complications related to this change of representation. There are many cases where a displacement and boosting operation:

$$\chi \rightarrow \bar{\chi} = \exp\ [i(k_\alpha R_\alpha - s_\alpha P_\alpha)]\ \chi, \tag{2.2}$$

just transforms χ into an other wave packet $\bar{\chi}$ whose structure is as simple as that of χ. The parameters k_α, s_α which label $\bar{\chi}$ actually provide all the necessary physical information to specify initial and final conditions of experiments. Intermediate states, of course, can as well be described in terms of $\bar{\chi}$.

Last but not least, wave packets remain in the Hilbert space. The aim of a theory being to provide finite numbers, the use of flexible and bounded wave functions is likely to help improving the accuracy and convergence of numerical calculations. As will be seen in the next section, there is no major difference between collision (continuum) problems and bound state problems, for a wave packet remains square integrable while it goes to infinity or remains at a finite distance.

3. Variational approach

All the information demanded by a theory of an atomic or nuclear system may be recovered if the matrix elements of the Green's function are known. Let $\mathcal{H}$ be the hamiltonian of the N-body system. A collection of matrix elements:

$$\mathcal{D} = \langle\chi'|(E + i\Gamma -\mathcal{H})^{-1}|\chi\rangle, \tag{3.1}$$

or in a slightly more complicated formalism, of matrix elements:

$$\mathcal{T} = \langle\chi'|V'(E + i\Gamma -\mathcal{H})^{-1}V|\chi\rangle, \tag{3.2}$$

obviously provide the Fourier transform of the evolution operator, or even more explicitly the off-shell T-matrix. Here χ and χ' are many product-type wave packet such as shown in eq. (2.1). They could be of the type shown by eq. (2.2) as well. If R_α is a distance between centers-of-mass, and P_α the corresponding momentum, $\bar{\chi}$ is furthermore an excellent approximation to a channel wave function (a boosted shell model).

In eqs. (3.1) and (3.2) an imaginary part $i\Gamma$ is added to the energy E in order to allow an on-shell limit $\Gamma \to 0$.. When E is in a range of bound states, poles of $\mathcal{D}$ or $\mathcal{T}$ give information on bound states. When E is in the continuum, one deals with a theory of collisions, obviously (in eq. (3.2) V and V' are prior and post potentials). All told the inversion of $(E+i\Gamma-\mathcal{H})$ is the only technical problem to face.

As long as Γ is finite the Green's function acting on a wave packet χ (or $V\chi$) gives again a wave packet. Consider the functional:

$$F \equiv \langle\Phi'|\chi\rangle + \langle\chi'|\Phi\rangle - \langle\Phi'|(E+i\Gamma-\mathcal{H})|\Phi\rangle, \tag{3.3}$$

where Φ and Φ' are infintely flexible wave packets. Variation and stationarity of F with respect to Φ and Φ' prove, trivially, that the stationary value of F is nothing but $\mathcal{D}$. An identical variational principle is obtained for $\mathcal{T}$ if V and V' are inserted in F.

The Green's function between wave packets can thus be calculated variationally with just wave-packet trial functions. The on-shell limit can be taken later.

In the next section we show how mean-field approximations, very convenient for practical calculations, can be implemented in this wave packet scheme.

4. Mean fields

Nothing prevents to restrict Φ, Φ' in eq. (3.3) to factorizable wave packets analogous to χ and $\bar{\chi}$. Then Φ and Φ' are no more infinitely flexible, but the resulting approximations $\bar{\mathcal{D}}$ to $\mathcal{D}$ (or $\bar{\mathcal{T}}$ to $\mathcal{T}$) are nonetheless of great interest.

This is because F contains a term (the last one) which is close to the Rayleigh-Ritz functional. Hence the ansatz:

$$\Phi = \mathcal{A} \prod_i \phi_i, \tag{4.1a}$$

$$\Phi' = \mathcal{A} \prod_i \phi'_i, \tag{4.1b}$$

generates from F a simple generalization of Hartree-Fock equations. A slightly detailed argument shows that one must solve equations of the form:

$$(\eta_i - h)\phi_i = S_i,$$
(4.2a)

$$\phi_i'(\eta_i - h) = S_i',$$
(4.2b)

where h is the usual Hartree-Fock hamiltonian and S_i, S_i' are source terms derived for the first two terms of F. Preliminary results show that this approximation can be excellent. For more details the reader is referred to [1] and [2].

5. Conclusion

Time independent wave packets are of major interest for microscopic calculations of the Green's function, for all the many-body integrals can be made factorizable. Both the initial and final conditions are kept under control by the variational method we have introduced. This contrasts with time dependent methods, where usually numerical approximations are nonlinear and thus incompatible with the linear superposition of channels.

It is a pleasure to thank the organizers of this Symposium for this opportunity of stimulating and encouraging exchanges.

References

1) B.G. Giraud, M.A. Nagarajan and I.J. Thompson, Ann. of Phys. (N.Y.) 152 (1984) 475.
2) Y. Abe and B.G. Giraud, Nucl. phys. A440 (1985) 311

QUANTAL ONE-BODY DISSIPATION
AND THE WALL FORMULA

J.J. Griffin

Department of Physics and Astronomy, University of Maryland,
College Park, Maryland 20742, USA

M. Dworzecka

Department of Physics, George Mason University,
Fairfax, Virginia 22030, USA

For a certain specially prescribed RPA approximate model problem whose quantal dissipation rate reduces precisely to the classical Wall Formula, we show that alteration of three of its idealizations towards nuclear realism reduces the quantal dissipation rate substantially. Thus the classical Wall Formula dissipation is placed into its proper relationship with the more realistic quantal descriptions of dissipation as a simplification which systematically and substantially overestimates the actual one-body dissipation rate.

The time-dependent RPA approximate solution of the Schrödinger equation for a certain model problem [1-6] is shown to yield in closed form precisely the collective energy dissipation rate of the classical Swiatecki Wall Formula [7,8], but only when certain unrealistic conditions are assumed. Altering three of these conditions towards nuclear realism reduces the quantal dissipation rate substantially from the Wall Formula value.

The most drastic reduction (to zero!) occurs when the quantal condition that the collective phonon energy must exceed the nucleonic binding energy is imposed. It reflects simply the fact that in the self-consistent RPA treatment of the collective wall motion, the two-body interactions are completely diagonalized among the collective phonon state and all the bound RPA (1p-1h) eigenstates. Since the Wall Formula incorporates no self-consistency between the wall and the particles, it cannot reflect this effect of the self-consistency.

The second largest reduction occurs when the nuclear potential is allowed to transmit, as well as to reflect, interior nucleons. (The classical Wall Formula "wall" is 100% reflecting, by specific assumption.)
This improvement reduces the dissipation rate by about one order of magnitude for typical nuclear situations.

Finally, the relaxation of the (implicit) classical assumption that the collective phonon energy, $\hbar\omega$, be negligible small leads to a smaller reduction of the dissipation rate, by a factor ~ 0.7 for typical giant resonances.

Thus seemingly innocent assumptions in the classical treatment may obscure serious physical simplifications. Also, since the removal of each such simplification is here shown to reduce the calculated dissipation rate, the classical Wall Formula appears to provide a maximal overestimate of the quantal one-body dissipation rate.

On the other hand, the inclusion of two-body dissipation (omitted by assumption both here and in the Wall Formula) will increase the overall dissipation rate. Thus this correction will aggravate Wall Formula's large overestimate of the dissipation rate, while ameliorating the too-small value of the present quantal one-body model.

In a continuing series of papers [1-6], C. Yannouleas et al. have provided a time dependent description of the decay of an RPA collective state (diagonal at time t=0 in a certain subspace, S_R) into the states of an additional subspace/S_A by residual two-body coupling. The method involves (a) diagonalization (in the RPA approximation) of the Hamiltonian in the space, S_R, of all bound 1p-1h states to determine the coherent linear combination, Q^+, of 1p-1h states which defines the collective one-phonon eigenstate; (b) construction of the collective state, $|c>$, as an arbitrary linear combination of pure phonon states to specify the time-dependent collective solution at t=0 (here we consider the exponential form of the packet used in ref. [2,3]; (c) solution of the time-dependent Schrödinger equation (also in the RPA) in the complete space, $S_R + S_A$, to determine the decay of the amplitude, $<c|\Psi(t)>$, of finding the system in the original collective state.

To extract the quantum analog of classical one-body dissipation we consider only the one-body part of the residual interaction, which connects the collective state only to states in S_A of the 1p-1h type. References [1-4] show that when the typical coupling matrix element, κ, and the energy spacing, d, of states in S_A both vary slowly [†], closed form approximate solutions can be obtained for the tipe-dependent RPA-Schrödinger equation [††]. Furthermore, that these simplify [1] when the coupling is "weak" to forms which are monotonically dissipative [*] for the first half of the Poincaré period [**] when the space S_A is "dense", i.e., when

---

† In fact, we assume the uniformly spaced "picket-fence" model for the spacing of the eigenenergies: $\hbar\omega_\nu = \hbar\omega_0 + \nu d$, where $\nu = 0, \pm1, \pm2, \pm3, \ldots$

†† See especially Appendix C of ref [1].

* Such behavior justifies the use of dissipative terminology, and even of a time irreversible approximation [3,4] to the (always time-reversible) Schrödinger equation.

** With the picket-fence level density (Refs. [1,10], the system is actually periodic in time with a Poincaré period, $T_p = 2\pi\hbar/d$.

the spacing between levels is much less than the coupling matrix element.

In addition to these qualitative results, Yannouleas' analysis[1-4]
yields an expression for the phonon decay width, as follows:

$$\Gamma = (2\pi/\hbar) \iint dp \ dh \ |K(p,h;\omega_c)|^2 \ \delta[\hbar\omega_c - \varepsilon(ph)].$$ (1)

Here the integral denotes sums over all of the labels of the particle-hole
states of S_A with energy $\varepsilon(p,h)$, $\hbar\omega_c$ is the energy of the collective phonon,
and $K(p,h;\omega_c)$ is the RPA matrix element for a phonon to annihilate into the
particle hole state, (ph).

We define the restricted subspace S_R, in which the RPA collective phonon
is diagonalized, to include the Fermi gas ground state (with Fermi energy, ε_F)
and all of the "bound" one particle one hole excitations. Here the word "bound"
implies that the particle state energy is less than $\varepsilon_F + B$, where B is the assumed
nuclear binding energy. Then the additional subspace S_A comprises all particle-
hole states with the particle in an "unbound" state, $\varepsilon_p > \varepsilon_F + B$. Note that, for
the fully reflecting nuclear potential when the depth of the nuclear mean field,
U_0, becomes much larger than $\varepsilon_F + B$ for $R > R_N$, as in our calculation of γ_{RP}^∞, the
separation between the spaces S_R and S_A still occurs at the particle energy
$\varepsilon_F + B$. Thus 1p-1h states with $\varepsilon_p > \varepsilon_F + B$ are defined to lie in S_A, and referred
to as "unbound", even though for $U_0 \to \infty$ their wave functions vanish at and outside
of the nuclear radius. Later, when finite values of U_0 are considered in a
Feynman box of infinite radius, U_0 will be set equal to $\varepsilon_F + B$ and the particle
states in S_A will have the form of truly "unbound" continuum states whose wave
functions extend over the infinite Feynman box.

For a wave packet which explicitly describes a time-varying small amplitude
multipole shape oscillation of the nuclear potential, the collective energy
dissipation rate is proportional to the kinetic part of the collective energy [3,4].
Then one finds

$$\dot{E}_{RP} = -2\Gamma[<p_c>^2/2M_c] = \gamma_{RP}(\dot{\alpha}_\lambda)^2$$ (2)

as the closest quantal analog of the classical Wall Formula result [7,8],

$$\dot{E}_{WF} = [\rho\bar{v}R^4](\dot{\alpha}_\lambda)^2 = [\hbar(k_F R)^4/8\pi^2](\dot{\alpha}_\lambda)^2 = \gamma_{WF}(\dot{\alpha}_\lambda)^2.$$ (3)

By simply calculating Γ from (1) and inserting it into (2) one can obtain the quantal RPA analog, γ_{RP}, of the Wall Formula coefficient, γ_{WF}. We note that the need to use the time-dependent packet in order to give the quantal dissipation rate (2) its close structural relationship to the classical Wall formula (3) is the first of several ways in which the quantal problem must be properly structured to reproduce every physical feature of the classical model, if the precise classical result is to emerge from the quantal calculation.

In general, when the RPA collective phonon describes the same multipole shape oscillation as the parameter $\alpha(t)$ in the Wall Formula (3), the quantal problem is as close as it can be to the classical problem even though the RPA mean field is in general nonlocal. But when the coupling interaction is chosen to be a certain [9)] a separable λ-pole two-body interaction,

$$V^{\lambda}_{mjin} = -\chi/2 \sum_{\nu} D^{\lambda\nu}_{mi} D^{\lambda\nu}_{nj}, \tag{4a}$$

with matrix element factors defined by the radial derivative of the average mean field, as follows,

$$D^{\lambda\nu}_{mi} = -<m|r(dU_0(r)/dr)Y_{\lambda\nu}(\theta)|i>, \tag{4b}$$

then the exponential packet of RPA phonons describes a mean field which not only oscillates in a pure λ-pole isoscalar shape vibration, but is also a local potential. The resulting model is referred to as the vibrating potential model [9)].

We utilize such a force to evaluate Γ in (1) and γ_{RP} in (2) for the case of a spherical square well equilibrium mean field of depth, $-U_0$. Before the RPA diagonalization, the particles fill the states up to the Fermi energy ε_F, as shown in Fig. 1. For small multipole order, λ, and large well-depth, $U_0 \gg \varepsilon_F + \hbar\omega_c \gtrsim \varepsilon_F + B$, the sums over the particle and the hole labels in (1) may be approximated by integrals which can be executed entirely in closed form (as detailed in Ref. [10)]) in terms of the following dimensionless phonon energy, and binding energy (or potential depth),

$$\eta = \hbar\omega/\varepsilon_F \quad \text{and} \quad \beta = B/\varepsilon_F \quad (\text{or } \zeta = (\varepsilon_F+B)/\varepsilon_F = 1+\beta). \tag{5}$$

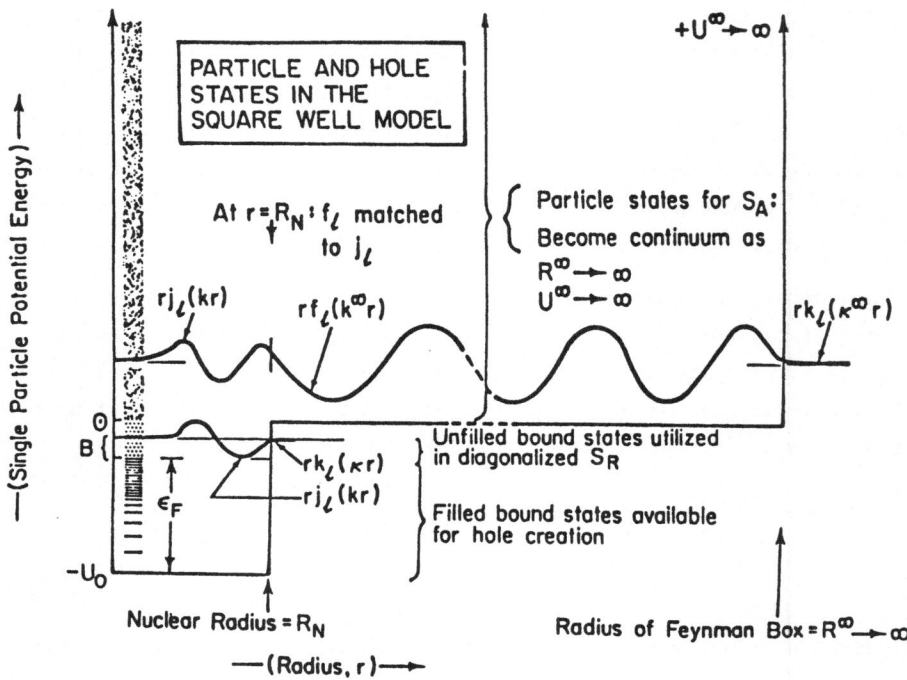

<u>Figure 1</u>: The finite well potential contained in a Feynman box is illustrated
together with typical bound and continuum state nucleonic wave
functions. The classical Wall Formula for dissipation is recovered
only in the limit when B=0 and $R^{\infty}=R_N$.

Then integration of (1) yields

$$\gamma_{RP}^{\infty} = (\frac{\hbar(k_F R_N)^4}{8\pi^2}) \{\frac{1}{3\eta} (1+\eta)^{3/2} + \frac{1}{4} \eta(1+\eta)^{1/2} + \frac{\eta}{4} (2+\eta)\ln[(\sqrt{1+\eta} - 1)/\sqrt{\eta}]$$

$$- \frac{1}{3\eta} \zeta^{3/2}(\zeta-\eta)^{3/2} - \frac{1}{4} \eta \zeta^{1/2}(\zeta-\eta)^{1/2} - \frac{\eta}{4} (2\zeta-\eta)\ln[(\sqrt{\zeta} - \sqrt{\zeta-\eta})/\sqrt{\eta}]\}$$

$$\text{for } \beta < \eta \leq \zeta \qquad (6)$$

and a similar expression for $\eta > \zeta$, obtained from (6) by replacing ζ everywhere
by η thereby deleting the last three terms.

For $\eta < \beta$ (= ζ-1), both Γ and γ_{RP}^{∞} vanish identically because no 1p-1h state
in S_A can honor the energy δ-function in (1). The dissipation rate (6) is plott
as a function of phonon energy, η, in Fig. 2 as a ratio to γ_{WF} of (3) for
various values of β.

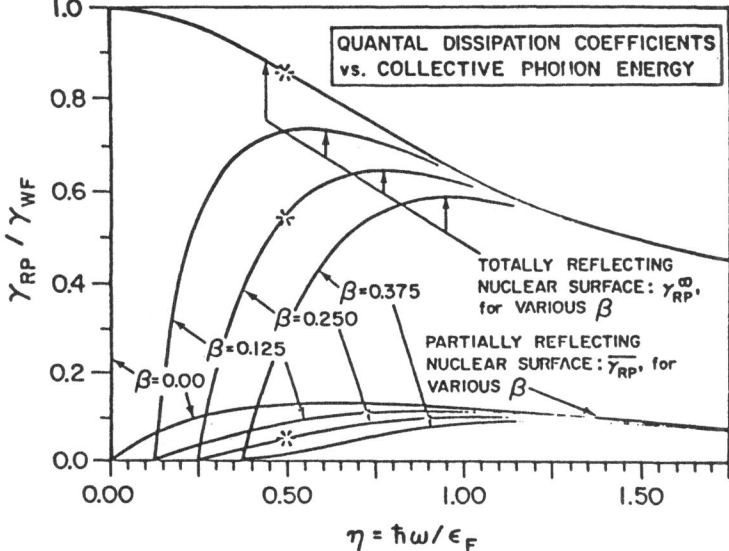

<u>Figure 2:</u> Various quantal dissipation rates are compared with that of the classical Wall Formula, γ_{WF}. (a) In every case, the quantal dissipation rate vanishes identically whenever the phonon energy is insufficient to excite a state of S_A; i.e., whenever $\hbar\omega < B$ (or $\eta < \beta$). The maximum rate γ_{RP}^{∞} for the totally reflecting nucleon field analogous to the classical wall is less than γ_{WF} for $\eta > 0$, but only by a factor of ≈ 0.7 for typical nuclear giant resonance situations, indicated by asterisks. (c) A much larger reduction to γ_{RP} ($\sim \gamma_{RP}/10$) results from the replacement of the infinitely deep mean field in (b) by a more realistic nuclear mean field which only partially reflects unbound nucleons.

A crucial result of the present analysis is that the limit $\hbar\omega \to 0$ of γ^{∞} in (6) is precisely the coefficient γ_{WF} of (3),

$$\gamma_{RP}^{\infty} \underset{\eta \to 0}{\to} (\hbar(k_F R_N)^4/8\pi^2) = \gamma_{WF}, \qquad (7)$$

when the binding energy, β, has been chosen equal to zero corresponding to $\zeta=1$. [†]

[†] We note that the Wall Formula has previously been extracted [2,4] from the $\hbar\omega \to 0$ limit of the frequency dependent dissipation of a system for which the additional subspace S_A, consisted of the complete set of bound 1p-1h states. We observe that this zero binding energy case is a very special and not very realistic situation in which the RPA subspace, S_{R},[2] is trivialized to a single state. We note that the expression given in Ref. [2], although it yields the correct $\eta \to 0$ limit, is erroneous. The correct, but not yet fully integrated, result for the special case of $\zeta=1$ was given in sect. VI.2 of Ref. [4]. However, the present generalization to $\zeta > 1$ is essential to exhibit the effects of self consistency within a nontrivial S_R subspace.

Thus an explicit, exact, closed form identity arises between the classical wall formula and this special case of the quantal RPA energy dissipation coefficient. Moreover, the special case is one which corresponds most precisely to the three specific physical assumptions of the Wall Formula in having (a) an infinitely deep square well bounding potential corresponding to the precisely localized fully-reflecting sharp "wall" itself; (b) a zero binding energy, $\beta=0$, which defines S_A to be the complete 1p-1h space, corresponding to the assumption of the non-self-consistent, externally driven wall; and (c) a phonon of energy $\hbar\omega \rightarrow 0$, corresponding to the infinite inertia required for the classical Wall to deliver arbitrary momentum and energy to the nucleons without recoil or slow down.

We note that even while it immuninates the specific nature of he Wall model assumptions, the larger quantal context underlines the fact that the Wall formula is defined by a very special selection of features which, if they seem natural enough in the classical context, are far less cogent in the more general quantum mechanical context. In particular, three of the physically unrealistic features of the classical model which are necessary to obtain the Wall Formula result from the quantal treatment can be altered towards nuclear realism while still allowing a simple analysis.

We first consider the phonon energy, $\eta = \hbar\omega/\varepsilon_F$. In fact Eq. (6) with the value of $\zeta = 1+\beta$ chosen equal to 1 (i.e., $\beta=0$) displays in Fig. 2 the effect of nonzero phonon energies. For a typical nuclear giant resonance situation, we take $\varepsilon_F = 40$ MeV, $\hbar\omega = 20$ MeV, so that $\eta = 0.5$. Then γ_{RP}^{∞} is reduced by a modest 15% from the classical value γ_{WF}, as indicated in Fig. 2.

The choice $\beta=0$ suppresses the effects of quantal self-consistency completely, since no 1p-1h excitation lies in the RPA space, S_R, where the self-consistent diagonalization was carried out. In such a quantal system even particles which gain a negligibly small energy from the wall can be excited by the residual interaction into the 1p-1h states of the additional subspace, S_A, just as they can in the classical externally driven Wall model. To exhibit the effect of quantal self-consistency, it is therefore essential to have a nontrivial particle-hole subspace included in S_R; i.e. , to have $\zeta>1.0$.

For illustration we choose $\zeta = 1.25$ equivalent to a binding energy of 10 MeV ($\beta = 0.25$). Then the dissipation rate γ_{RP}^{∞} in (6) is reduced by self-consistency identically to zero for phonon energies less than the binding energy ($\eta < 0.25$ in Fig. 2); for the typical nuclear giant resonance energy of $\eta = 0.5$, it is reduced below the $\beta = 0$ value again by a modest, but additional, 10 to 15%.

The third physically unrealistic aspect of the classical Wall formula is the assumption that the nuclear surface totally reflects nucleons, whatever their energy. This property has been incorporated into Eq. (6) by the assumption that all the particle states, even those with $\varepsilon_p > U_0$, vanish at and outside the nuclear radius R_N, and so correspond to solutions in the infinitely deep square well resulting when the Feynman box radius $R\infty$ in Fig. 1 is set equal to the nuclear radius R_N. This totally reflecting assumption can be relaxed by assuming the more realistic form of Fig. 1 for the nuclear mean field and allowing the "unbound" particle states with $\varepsilon_p > U_0$ to have the form of proper continuum states in the infinite Feynman box.

We have therefore calculated the averaged dissipation rate $\bar{\gamma}_{RP}$ for a square well of definite depth $U_0 = \varepsilon_F + B$. For such a potential the matrix elements (4b) of the derivative of the mean field at the nuclear radius are substantially reduced, introducing a reduction factor into the integrand of (1) which oscillates with the particle state energy, reflecting the continuum resonances in the shell model potential of Fig. 1. When these resonance oscillations are replaced by an average value, the expression (1) for the dissipation rate can be partially integrated in closed form, up to the integration over particle energy [10]. The result is

$$\bar{\gamma}_{RP} = (\frac{\hbar(k_F R_N)^4}{8\pi^2}) \; (\frac{4\zeta}{3\eta}) \int_{z_G}^{1+\eta} dz \; [(z-\eta)^{3/2}(z-\zeta)]/[(2z-\zeta)z^{1/2}], \qquad (9)$$

where $z = \varepsilon_p/\varepsilon_F$ is the dimensionless particle energy, and the lower limit, z_G ($\leq 1+\eta$), is given by the greater of ζ and η and defines the result for the separate ranges, $\beta < \eta < \zeta$ and $\eta > \zeta > \beta$, respectively.

Numerical integration of the expression (9) for the rate, $\bar{\gamma}_{RP}$, yields the results plotted in Fig. 2 against the dimensionless phonon energy, η, for various binding energies, $\beta = \zeta - 1$. Note that the rate (9) also vanishes whenever $\eta < \beta$, again reflecting the quantal self-consistency. For the typical nuclear giant resonance mentioned above, $\zeta = 1.25$ and $\eta = 0.5$, and the dissipation rate $\bar{\gamma}_{RP}$ is about 9 × less than γ_{RP}^{∞}. This factor reflects the error generated by the Wall Formula assumption of a totally reflecting nuclear potential.

In fact, Fig. 2 summarizes all of the major results of this study insofar as the relationship of the quantal γ_{RP} to the classical γ_{WF} is concerned: the Wall Formula always overestimates the realistic quantal one-body dissipation rate, $\bar{\gamma}_{RP}$. Nor can it be claimed that the quantal approximations have omitted some important physics which the Wall Formula includes, since precisely the Wall Formula result is obtained when every option is exercised specifically to make the quantal case most resemble the specific classical assumptions of the classical Wall model. We learn thereby how important are the implications of these seemingly innocent classical assumptions: independence of wall and particles, infinite reflectivity of nuclear potential, and small phonon energy.

Finally, we note that the complete quantal treatment must include [3,4] additional dissipation of probability flux into 2p-2h states, which is not considered in the present analysis of the one-body part of the dissipation. The magitude of such 2p-2h contributions to the width of the giant dipole resonance was long ago [11] estimated by a microscopic perturbation theoretic stationary-state treatment and found to yield the dominant part of the width for heavier nuclei. More recent analysis of the isoscalar E2 resonances show [12] the one-body dissipation to be a small part of the obseved width [13,14], even for light nuclei [15], whereas coupling to 2p-2h states provides the main contribution to the damping [15-18]. The present analysis of one-body dissipation is in qualitative agreement with these microscopic results.

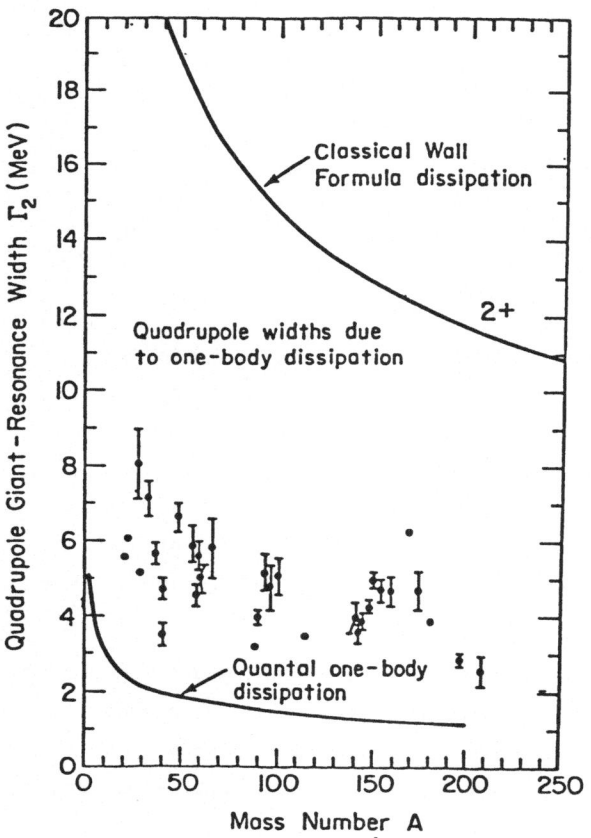

Fig. 3: (This figure is adapted from Ref. [19] with the kind permission of the authors.) The 2+ widths calculated from the classical Wall Formula [19] one-body dissipation rate are plotted versus mass number, A, and compared with the data of Refs. [13,14]. Also indicated are the widths implied by a more realistic quantal dissipation rate one-tenth as large as that of the Wall Formula, which corresponds roughly with the maxima of the dissipation rates, $\overline{\gamma_{RP}}$, for the partially reflecting quantal cases of Fig. 2, and therefore provides still an upper limit on the true one-body dissipation, denoted by the asterisk at $\overline{\gamma_{RP}} \approx 0.05 \, \gamma_{WF}$ for $\eta = 0.50$.

Thus one is led to the conclusion that the Wall Formula's (~3x) overestimate [19] of the experimental widths [13,14] of giant quadrupole resonances, as
summarized in Fig. 3, is due to the unrealistic idealizations of the classical
Wall model. The more realistic quantal estimate given by $\bar{\gamma}_{RP}$ of Eq. (9) is
some 10× smaller than the Wall Formula value, and therefore about 3× smaller than
the observed widths. This result is illustrated in Fig. 3 by the curve labelled
"Quantal One-Body Dissipation". Then the residual discrepancy between the too
small quantal one-body dissipation and the observed widths seem very likely
to be explainable in terms of the dissipation to 2p-2h states in S_A, which is
omitted from both the classical Wall Model and from the present quantal one-
body treatment.

The present situation poses a subtle ambiguity for nuclear phenomenology,
in that calculations based upon the Wall Formula might yield good results not
because the classical Wall Formula is correct, but (e.g.) because both 1p-1h
and 2p-2h dissipation occur largely in the nuclear surface (where the Wall
Formula calculations perhaps correctly places its dissipation), and because
the Wall Formula's incorrectly large value of the one-body dissipation rate
artificially and accidently compensates for its omission of the 2p-2h dissipation. To resolve such an ambiguity, one must seek to match experimental and
theoretical signatures which are able specifically to distinguish the one-body
from the two-body dissipation, and to avoid a false security based simply on
comparisons of undifferentiated total dissipation rates. In the meanwhile,
a Comprehensive (one-body plus two-body) Surface Dissipation might reasonably
be considered for phenomenological use, which has the same form as the Wall
Formula dissipation, but a strength which is adjected to the total widths of
giant resonances rather than to the simple, but unrealistic, classical Wall
model.

Support of the U.S. Department of Energy for this research is gratefully
acknowledged, as well as the generosity of Drs. J.R. Nix and A.J. Sierk in
allowing the use of their published figure as a basis for our Fig. 3, and the
hospitality of the Lewes Center for Physics of the University of Delaware,
which one of us (JJG) enjoyed during some of this work. One of us (MD) wishes
also to thank the University of Maryland Nuclear Theory Group for their
hospitality during her frequent visits there.

References

1) C. Yannouleas, M. Dworzecka and J.J. Griffin, Nucl. Phys. A379 (1982) 256.

2) C. Yannouleas, M. Dworzecka and J.J. Griffin, Proc. Xth Int. Workshop on Gross Properties of Nuclei, Hirschegg, Austria, January 1982, ed. by H. Feldmeier (Inst. f. Kernphysik, T.H. Darmstadt, 1982) p. 163; Proc. Workshop on Nuclear Dynamics, Granlibakken, Tahoe City, CA February 1982 (U. of California, Berkeley, Conference Report #LBL14138, 1982) p.40.

3) C. Yannouleas, M. Dworzecka and J.J. Griffin, Nucl. Phys. A397 (1983) 139.

4) C. Yannouleas, U. of Maryland, Ph.D. Dissertation, ORO #5126-158 (July, 1982).

5) C. Yannouleas, Preprint GSI-83-25, private communication.

6) J.J. Griffin and M. Dworzecka, Proc. Int. Sympsium on Nuclear Excited States, Lodz, Poland, June 1984, ed. by M. Przytula, to be published.

7) D.H.E. Gross, Nucl. Phys. A240 (1975) 472.

8) J. Blocki et al., Ann. Phys. 113 (1978) 330.

9) D.J. Rowe, Nuclear Collective Motion, Sec. 16.7 (Methuen, London, 1970).

10) J.J. Griffin et al., U. of Maryland ORO #(5126-146, to be published.

11) M. Danow and W. Greiner, Phys. Rev. 138B (1965) 876.

12) S. Shlomo and G. Bertsch, Nucl. Phys. A243 (1975) 507.

13) G.R. Satchler, in Proceedings of the International School of Physics "Enrico Fermi", Varenna, Italy 1976, Course LXIX, Elementary Modes of Excitation in Nuclei (Societa Italiana di Fisica, Bologna, 1977) p. 271.

14) F.E. Bertrand, Ann. Rev. Nucl. Sci. 26 (1976) 457.

15) J.S. Dehesa et al., Phys. Rev. C15 (1977) 1858.

16) G.F. Bertsch et al., Rev. Mod. Phys. 55 (1983) 287.

17) V.G. Soloviev et al., Nucl. Phys. A288 (1977) 276.

18) P.F. Bortignon and R. Broglia, Nucl. Phys. A371 (1981) 405.

19) J.R. Nix and A.J. Sierk, Phys. Rev. C21 (1980) 396; C25 (1982) 1068.

LOCALIZATION OF SINGLE-PARTICLE
STATES IN THE CONTINUUM

Y. Hahn

Physics Department, University of Connecticut, Storrs, CT 06268, USA

Many particle scattering theory attempts to extract various solutions
for different physically relevant boundary conditions. The richness of the
system however makes this task difficult. One natural approach would be to
extend the shell model or the HF method, which have been successful for
bound state problems. In fact, the shell model was extended [1] to the continuum problem (continuum shell model, CSM) nearly twenty years ago, in which a
single nucleon continuum orbital was included in the determinantal functions,
and the residual interaction was taken into account by solving a set of coupled
equations. Some of the numerical complications were elaborated on in ref. [1].
The method had some limitations; in particular, the model cannot accomodate
more than one particles in the continuum, so that the reactions such as (p,d)
and (t,a) are excluded. An attempt to resolve this problem was made earlier [2],
and the problem was reexamined [3] in 1980. An alternative to CSM is the TDHF
procedure, in which the time-dependent determinantal function is used with
self-consistency requirement. This mean-field approach (MFA) has been modified
to extract collective variables [4,5], especially for large-amplitude but
small velocity modes. Presumably, the generator coordinate method (GCM) can be
used also to treat this problem [6,2]. In practice, however, it is rather
difficult to carry out such programs, except in some simple cases. We consider
here a simple formulation of the problem within the framework of the CSM,
restricting ourselves to the case of two particles in the continuum.

The localization of two particles in the continuum is realized here by an
explicit construction of a wave packet, thus at the same time making the wave
function square-integrable. The width of the packet is large in coordinate
space. To simplify the disussion, we assume that the target is very heavy with
A fermions, A >> 2, so that the Center-of-mass problem for the target is neglected. On the other hand, the center-of-mass motion of the beam particle with
two fermions should be properly taken into account, using the GCM projection
procedure. (Later, we consider a simpler approach, by passing the need for
the projection.) To construct a set of localized single-particle orbitals
to be used in the CSM for the reaction

$$B = (1+2) + (A) \;\to\; B + A$$
$$\to\; 2 + (1+A) \;, \text{ etc.}$$

we consider a model Hamiltonian for the A+2 particles

$$H^M = \sum_{i=1}^{A+2} h_i^M \equiv H - V^M, \tag{1}$$

where

$$h_i^M(x_i) = -\frac{\nabla x_i^2}{2m_i} + V_A^M(x_i) + v_B^M(x_i, R') \equiv h_i^M(x_i, R') \tag{2}$$

and the model potentials are of the form (of molecular type)

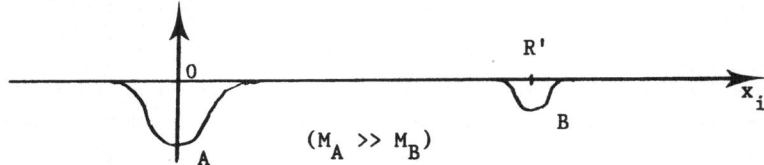

In (2), R' is a parameter (generator coordinate). Then

$$h_i^M(x_i)\ \phi_n^M(x_i, R') = e_n^M(R')\ \phi_n^M(x_i, R'). \tag{3}$$

These molecular states can now be used to construct the cluster functions for B

$$\phi_\alpha^B(1,2) = \frac{1}{\sqrt{2}}\ \det\ |\phi_{n1}^B(x_1, R')\ \phi_{n2}^B(x_2, R')| \equiv \Phi_\alpha^B(x_1, x_2; R') \tag{4}$$

in which only those orbitals ϕ_n^B, which are localized in the region of v_B^M, are included. Similarly, we also have for the target cluster A

$$\phi_\alpha^A(1,\ldots A) = \frac{1}{\sqrt{A!}}\ \det\ |\phi_{n1}^A(x, R')\ldots\phi_{nA}^A(x_A, R')|, \tag{5}$$

which contain only those orbitals in the region of V_A^M and are obviously <u>mildly</u> dependent on R'. The total channel function is then given by

$$\phi_\alpha^M = \mathbf{A}_{A,B} \{\phi_\alpha^A \phi_\alpha^B\}, \qquad (6)$$

where $\mathbf{A}$ describes the ansitymmetrization of particles between the clusters A and B, with the correct normalization factor $[2.A!/(A+2)!]^{1/2}$.
Using the usual Peierls-Thouless(PT) projection procedure [6], as also discussed for example in refs. 2 and 3, we have the purely internal wave functions of zero relative kinetic energy

$$\phi_{\alpha 0}^{PT} = \int dR' \ F_\alpha(R-R') \ \phi_\alpha^M(x,R-R',R'). \qquad (7)$$

Obviously, x_i for those particles in A are not affected $(x_i=r_{Ai})$, while those associated with B will be $x_i=r_{Bi}+R$, where, for the equal mass case, $R=(x_i+x_j)/2$. The plane wave of the problem is then

$$\phi_{\alpha K}^{PT} = e^{iK.R} \ \phi_{\alpha 0}^{PT}. \qquad (8)$$

We now construct a wave packet of the form

$$\phi_{\alpha K}^{WP2} = \int dK' \ \tilde{a}_\alpha^{(2)}(K'-K) \ e^{iK'.R} \ \phi_{\alpha 0}^{PT}, \qquad (9)$$

for (8). This, together with a similar form for the scattered part, can be used to describe the A+B scattering.
However, instead of the general form (9), which is applicable to more complex cluster geometries, we write directly for the two-particle B

$$\phi_{\alpha K}^{(2)} = \mathbf{A}_{AB}\{\psi_\alpha^A(3,\dots,A+2) \int dK_1' \int dK_2' \ \tilde{a}_\alpha^B(K_1'+K_2'-K)\tilde{\psi}_\alpha^B(K_1'\mu_2-K_2'\mu_1)e^{iK_1'x_1+iK_2'x_2}\} \quad (10)$$

where in general $R_{12} = (m_1x_1 + m_2x_2)/(m_1+m_2)$, $\mu_1 = m_1/(m_1+m_2)$, $\mu_2 = m_2/(m_1+m_2)$. It is important to note that the cluster structure of B = (1+2) in (10) is maintained through <u>both</u> the $\tilde{\psi}_\alpha^B$ and $\tilde{a}_\alpha^B$; they are essential in deriving (10), which is in the 'single-particle' form and still square-integrable. The usefulness of (10) cannot be stressed enough. Together with the wave function containing one single continuum orbitals

$$\Phi_{\beta K}^{(1)} = \mathbf{A}_{A+1,1} \{\psi_{\beta}^c(2,\ldots A+2) \int dK_1' \; \tilde{a}_{\beta}^{(1)}(K_1'-K) \; e^{iK_1'x_1}\}, \qquad (11)$$

we have the necessary set of functions to construct a scattering theory of interest here.

An example: We apply the wave packet formalism of CSM with (10) to the following reaction

$$\begin{array}{c} H + Ca^{17+} \rightarrow p + (Ca^{16+})^{**} \rightarrow (Ca^{16+})^* + p + \gamma, \qquad (12) \\ (1s) \quad (1s^2 2s) \quad (1s2s2pn\ell) \\ B \qquad A \end{array}$$

which has been studied recently at Berkeley and at Brookhaven [8]. It is closely related to the collision process (dielectronic revombination)

$$e + Ca^{17+} \rightarrow (Ca^{16+})^{**} \rightarrow (Ca^{16+})^* + \gamma, \qquad (13)$$

which is known to be an important process in astrophysical and laboratory plasmas. Comparing (12) and (13), the beam B in (12) is effectively providing an "electron beam" for the process (13). So far, direct experiment of (13) has been proven difficult, and only (12) has been possible; in the actual experiment, the 'target' A was accelerated and collided with the beam gas of H_2, He, Ar etc. The theoretical treatment of the process is given [9] in terms of the wave functions (11), which simplifies considerably in the case of $m_e = m_1$ and $m_p = m_2$ with $m_2 \gg m_1$. That is,

$$\Phi_{\alpha K}^{(2)} \approx \int dK_1' \; \tilde{\psi}_{\alpha}(K_1 - k_2\mu_1) \; \tilde{a}_{\alpha}(K_1 - k_1) \; e^{iK_1'x_1 + iK_2'x_2}, \qquad (14)$$

where $\exp(iK_1'x_1)$ is the all-important factor (electron translation factor). This factor is required to correctly satisfy the asymptotic boundary condition, and thus played a crucial role in the usual perturbed stationary state method for ion-atom collisions with charge exchanges [10,9]. Aside from the wave-packet factor $\tilde{a}_{\alpha}$ and the Compton profile $\tilde{\psi}_{\alpha}$ in (14), $\exp(iK_1'x_1)$ also shows that physically there is an incoming electron beam in this problem on the target Ca^{17+}. The amplitude for the process (12) is given by

$$T^{RTE} \approx \sum_{d} \int dK_1' \tilde{\psi}_\alpha (K_1' - k_2 \mu_1) \tilde{a}_\alpha (K_1' - k_1) \ [<\Psi_{A'} (1_a, 3_a); \vec{k}_j |D|$$

$$\cdot \phi_{A'} (1_a, 3_a); \vec{o}_j > \frac{1}{e_{a1} + e_{a3} - E_{d'} + iP(d)/2} <\phi_{A'} (1_a 3_a) |V_{13}| \psi_A (3_a) e^{iK_1' x_1} >] \qquad (15)$$

where D is the dipole photon-electron coupling and V_{13} is the interaction between the electron 1 of B and electron 3 of A.

The experimental result agrees reasonably well with the theoretical cross section derived from (15). The charge state of Ca^{16+} and x-ray emitted by the captured ion were detected in coincidence.

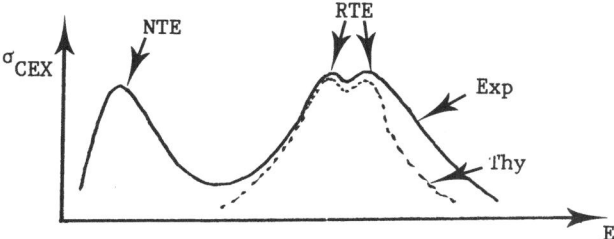

The low energy peak corresponds to the non-resonant charge exchange, in which the core proton in B plays a dominant role.

Conclusion

The proper wave packet construction for two particles in the continuum, given by (10), allows not only (a) a completely antisymmetric formulation of the (A+2) particle collision problem, but also (b) allows an effective localization of the two particles in the continuum within the CSM. In addition, the electron translation factor is given correctly by (14) in this special case of $m_1 \ll m_2$. We have thus succesfully extended the CSM to the case of two-particle continuum, without the projection procedure. The requirement of the availability of $\tilde{\psi}_\alpha^B$ is not serious, because, to fully define the scattering problem with the A+B channels open, one has to supply ψ_α^B as a boundary condition in any case. More stringent test of the formalism discussed here is to apply it to a (d,p) reaction in which both (10) and (11) appear. A coherent CSM picture may be developed in terms of these functions. For more complicated clusters, the GCM

procedure outlined here and in refs. [2] and [3] may not be avoidable.
In the present treatment, the main simplification is brought about by the
assumption that A is very heavy and that the internal cluster functions are
available.
Finally, the form (10) would also be useful in formulating the mean field
approximation. This will be discussed elsewhere [9].

References

1) C. Bloch and V. Gillet, Phys. Lett. $\underline{16}$ (1965) 62 and $\underline{18}$ (1965) 58

2) Y. Hahn, Phys. Rev. $\underline{174}$ (1968) 1168; Nucl. Phys. $\underline{A146}$ (1970) 62;
   Ann. Phys. $\underline{58}$ (1970) 137

3) B. Giraud and Y. Hahn, Phys. Rev. $\underline{C23}$ (1981) 1486 and 1495

4) F. Villars, Nucl. Phys. $\underline{A285}$ (1977) 269

5) M. Baranger and M. Veneroni, Ann. Phys. $\underline{114}$ (1978) 123

6) R.E. Peierls and D.J. Thouless, Nucl. Phys. $\underline{38}$ (1962) 154; R.E. Peierls and
   J. Yoccoz, Proc. Phys. Soc. $\underline{A10}$ (1957) 381

7) Y. Hahn, Nucl. Phys. A, to be published.

8) J. Tanis et al., Phys. Rev. Lett (1985)

9) Y. Hahn, Phys. Rev. (to be published); also see J. Phys. $\underline{B11}$ (1978) 3221

10) D.R. Bates and R. Mc Carroll, Proc. Roy. Soc. $\underline{A245}$ (1958) 175

RECENT PROGRESS IN
SEMICLASSICAL WAVEPACKET METHODS FOR
CHEMICAL PHYSICS APPLICATIONS

E.J. Heller
University of Washington BG-10, Seattle WA 98195, U.S.A.

J.R. Reimers
University of California B014, San Diego, CA 92093, U.S.A.

Recent work in wavepacket dynamics in our group has been proceeding along two main branches. On the one hand, we have been developing new methods for wavepacket propagation appropriate to new areas of application, including inelastic scattering from large target systems, neutron scattering, finite temperature spectroscopy, semiclassical quantization and rotation-vibration spectroscopy. On the other hand, we have been applying our previously existing methods to a widening variety of complex problems, including the electronic spectra of polyenes and the spectrum of C_2H at 3000°K.

Of particular concern at this moment are three things. First, we continue to develop and expand our methods for semiclassical quantization using arbitrary trajectories. Using frozen gaussians following arbitrary trajectories, we use the spectral peaks in the fourier transform of an autocorrelation function to extract estimates of the eigenvalues for the potential surface under investigation. These eigenvalues can be shown to be directly related to the EBK eigenvalues of standard semiclassical quantization techniques. The EBK method requires a search for trajectories which satisfy three quantum additions in three dimensions, etc. Our "zero quantum condition" method can be shown to give rise to energies which are linearly related by extrapolation to the EBK energies.

Even more important than the energy eigenvalues are the wavefunctions. These are also obtained by Fourier transform of wavepackets which follow classical trajectories. We have recently made a breakthrough in understanding the reason why this technique works so well, when the actual wavepacket dynamics which is used as input is so poor. The reason it is poor is that the wavepackets should spread, and in fact become very nongaussian, and yet we use a nonspreading frozen gaussian as input to the dynamics. We have recently discovered that this frozen gaussian dynamics is close to solving a time-

dependent wave equation that is not a Schrödinger equation, but shares with
the Schrödinger equation the same eigenfunctions. This remarkable property
accounts in large part for the success of the semiclassical quantization
methods for wavefunctions. The method is called "Photo Spectral Quantization".

Recent progress has allowed us to treat the vibration-rotation problem
with wavepackets. Of longstanding interest has been the problem of pure
rotations. Recently, we have shown that a rotor in a plane can be solved
analytically by a superposition of gaussian wavepackets in symmetry adapted
coordinates. Here again, the frozen gaussians emerge as being remarkably
accurate, in fact exact for this problem. The hope is that for nonrigid
polyatomic molecules, the analogous techniques will give excellent vibration-
rotation eigenstates and eigenvalues. This hope is not yet confirmed, but
work is in progress.

Further work involving correlation functions is under way. A whole
range of spectroscopies can be obtained by directly calculating the underlying correlation function, which is then Fourier transformed to obtain this
measured spectrum. These correlation functions fall in two categories.
First, there are the quantum correlation functions, which have no classical
analog. Second, and much more commonly discussed, are the correlation
functions (such as that for infrared absorption) which have a classical analog.
Recently we have found that the usual interpretation of these correlation
functions in terms of "linear response" theory leads to some unnecessarily
poor semiclassical approximations.
For example, in the vibration rotation spectrum of a diatomic molecule at
finite temperature, one sees experimentally P and R-branched lines, which
are fine structure under the envelope of the P and R branches. In the
usual semiclassical approach to the dipole autocorrelation function for a
diatomic molecules at finite temperature, only the envelopes are obtained.
However, using wavepacket dynamics instead of straight classical trajectories,
one is able to recover the P and R branch structure under the bands. This
will permit a whole range of new applications involving collisional rotational broadening for one example.

New methods for semiclassical propagation are under investigation.
One of these methods involves coherent state path integrals, and the determination of the coherent state amplitudes of the propagator by full semiclassical root-search methods, which preserve Hermiticity and unitarity.
There is real hope that such new methods for semiclassical wavepacket
propagation will be of use in the treatment of small amplitudes (such as

tunneling), and for the treatment of longer time processes where single wavepacket dynamics methods break down.

In the United States, a whole range of wavepacket applications are now underway. In our group as well as in the group of Prof. Horia I Metiu at Santa Barbara, wavepacket dynamics as applied to atom-surface collisions in several kinds of situations has been or is being applied. Michael Herman (Tulane) and Andrew DePristo (Iowa) are working on methods of wavepacket propagation, atom-surface scattering (DePristo), nonadiabatic curve crossing (Herman), and other applications. Rex Skodje and Rob Coalson, both of whom attended the meeting, have developed a large array of useful applications and new techniques for wavepacket propagation. Bill Harter (Georgia Tech) has developed new wavepacket techniques for rotational motion in polyatomic molecules. Many experimentalists now interpret their work in terms of wave-packet dynamics.

Perhaps it can all be summed in a statement that "any process in quantum mechanical measurement can be stated and solved in terms of the dynamics of initially localized gaussian wavepackets". One can augment this statement with two facts. First, the use of such initially localized gaussian wave-packets has become a very efficient computational tool for attacking compli-cated dynamical problems numerically. Second, and even more important, the understanding of quantum phenomena in turns of localized wavepacket trajec-tories leads to a new intuitive tool, which has lead to deep insight into a number of spectroscopic and collisional processes of primary importance in chemical physics.

GAUGE INVARIANT PERIOD QUANTIZED WAVE PACKETS

Kit-Keung Kan

205 South Whiting Street, Alexandria, Virginia 22304, U.S.A.

1. Introduction

Wave packets are quantum mechanical entities which can accomodate classical concepts. In many ways time-dependent Hartree-Fock (TDHF) wave functions are wave packets. As can be seen in the numerical computations [1] of heavy ion collisions, the TDHF wave functions are localized in space, which translate, deform, vibrate, coalesce and breakup like a chunk of classical matter.

We envisage periodic motions in a classical description of collective vibrational and rotational states of a nucleus. Thus, we seek the periodic solutions in the TDHF description of such collective states. More precisely, we seek those periodic solutions which are gauge invariant, i.e., those independent of the choice of reference point of the energy. This is the basic motivation of the gauge invariant periodic quantization method [2-6].

In this paper we review the gauge invariant periodic quantization method in its general form, [4] summarize the exploratory examples in support of the method, and discuss its relationship with the exact stationary states in terms of time average.

2. Gauge invariant periodic quantization

The gauge-invariant periodic quantization (GIPQ) method is based on the time-dependent variational principle,

$$\delta \int_{t_0}^{t_1} \langle \Phi_c | (i\hbar\partial_t - H) | \Phi_c \rangle dt = 0, \tag{1}$$

where Φ_c is varied inside a manifold M_c in the Hilbert Space. This manifold is a direct product of a manifold M of wave functions of unit norm and the field of complex numbers C. The variation in C leads to the conservation of the norm and the determination of the time-evolution of the phase [2,4,7] of the complete wave function. With this phase Φ_c is given by

$$\Phi_c = \Phi \exp[i \int_{t_0}^{t} <\Phi|i\hbar\partial_{t'}-H|\Phi>dt'/\hbar]. \tag{2}$$

The variation in M determines Φ which contains the information of the specific dynamics of the physical system.

The energy $<H>$ of the system is conserved under the variational principle (1) and we can rewrite equation (2) as:

$$\Phi_c = \Phi \exp[i \int_{t_0}^{t} <\Phi|i\hbar\partial_{t'}|\Phi>dt'/\hbar].\exp(-i<H>t/\hbar)$$

$$\equiv \Phi_G \exp(-i<H>t/\hbar). \tag{3}$$

Here Φ_G is referred to as the gauge-invariant factor of Φ_c, since it is unaffected by gauge transformations of the form $H \rightarrow H+g(t)$, where $g(t)$ is an arbitrary function of time alone.

The GIPQ method seeks the time-dependent approximation to the eigenstates of H in which the gauge invariant wave function Φ_G is periodic. This is equivalent to selecting all the periodic solutions of Φ which obey the quantization rule:

$$\int_{0}^{T} <\Phi|i\hbar\partial_t|\Phi>dt = 2n\pi\hbar. \tag{4}$$

It is shown in ref.8 that the time-dependent variational principle is equivalent to Hamilton's principle in classical mechanics and that this quantization rule (4) can always be casted in Bohr-Sommerfeld's form.

The GIPQ wave function Φ_c in (3) is analogous to that of an exact stationary state which consists of an eigensolution and an energy phase. The time-independent, gauge-invariant eigensolution in the exact stationary state is replaced here by a periodic gauge-invariant solution.

3. Exploratory Applications of GIPQ

The GIPQ method has been applied to several exploratory one-body and many-body systems:

(i) Oscillating minimum wave packets in a simple harmonic oscillator (SHO) potential. GIPQ leads to the exact energy spectrum including the zero point energy [4,6].

(ii) Dilatational wave packets in a simple harmonic oscillator potential [6]. The dilatational wave packets are set up by initially stretching the minimum wave packet. Such wave packets all have even parity. The GIPQ rule then selects a set of wave packets with discrete initial amplitudes and reproduces the exact energy spectrum of the even-parity stationary states.

(iii) TDHF approximation of a two-particle system with SHO interaction in one dimension [2]. The single particle wave functions are dilatational minimum SHO wave packets. When the center-of-mass energy is removed, we obtain a spectrum of $1.5\hbar\omega(n+1/2)$ as compared to the exact spectrum of $\sqrt{2}\hbar\omega$ $(n+1/2)$.

(iv) TDHF Lipkin-Meshkov-Glick (LMG) model. Coherent single determinantal wave functions are considered [2] in solving the GIPQ problem for the LMG model. The result is illustrated in figure 1 where the energy levels as functions of the interaction strength χ for the case with A=50 is shown. It shows that the TDHF states undergo a transition from nondegenerated to doubly degenerated states when they cross the unperturbed ground state energy. Such a transition for the lowest level was previously studied as a phase transition by use of coherent states in the thermodynamical limit [9,10].

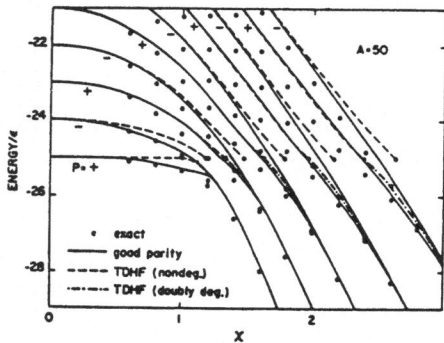

Figure 1: Comparison of exact, GIPQ good-parity projected TDHF, and GIPQ TDHF spectra as functions of the interaction strength χ for LMG model with A=50 particles.

(v) Projected TDHF LGM model. The coherent single-determinantal wave functions are projected to good-parity states before the GIPQ procedure is applied [3]. In this way the degeneracy in (iv) is removed and the new spectrum is qualitatively and quantitatively superior to that of (iv). See figure 1.

(vi) One dimensional collective rotation [5]. Under the assumptions that the rotation takes place about a fixed axis and that the many-body wave function is stationary in the body-fixed frame, the GIPQ method leads to the correct quantization of the rotational energy,

$$E^{rot} = I^2 \hbar^2 / 2 \mathcal{I}.$$ (5)

With Ω as the velocity of rotation, the moment of inertia,

$$\mathcal{I} = (m/\Omega^2) \int \vec{J} . \vec{\Omega} \times \vec{r} \, d\vec{r},$$ (6)

is identical in form to the one previously obtained in Schrödinger fluid [11] under a more restricted assumption. Here, $\vec{J}$ can be the total current of the nonperturbative many-body wave function.

Besides these exploratory applications, the GIPQ method has been used to calculate [12] breathing mode energies in doubly closed shell nuclei and obtained good agreement with experiments.

4. GIPQ exact Schrödinger states and their time average.

Since the GIPQ method is formulated on any manifold in the Hilbert space, it is also applicable in the full Hilbert space. In the full Hilbert space, the variational principle (1) leads to the exact Schrödinger equation. What is the relationship between the GIPQ and the exact eigenstates in the full Hilbert space ?

Any stationary state is a GIPQ solution in the full Hilbert space since a stationary state is gauge invariant and periodic (with period zero). However, not all GIPQ solutions are stationary states. The GIPQ oscillating and dilational wave packets in examples (i) and (ii) of Section 3 are solutions of the exact SHO Schrödinger equation, but they are not eigenstates.

For a general H, one can construct infinitely many non-stationary GIPQ solutions by making linear combinations with two stationary states. Let Ψ_1 and Ψ_2 be two exact eigenstates with energies E_1 and E_2, respectively. The following time-dependent wave functions,

$$\Phi_c = \{\sqrt{\kappa_1}\Psi_1 \, \exp[-i\kappa_2(E_1-E_2)t/\hbar] = \sqrt{\kappa_2}\Psi_2 \, \exp[-i\kappa_1(E_2-E_1)t/\hbar \,.\exp(-iEt/\hbar)] \quad (7)$$

where

$$\kappa_i = k_i/(k_1+k_2) \, , \qquad i=1,2 \text{ and integer } k_i \tag{8}$$

are GIPQ solutions [4,6] with energies given by

$$E = (k_1E_1+k_2E_2)/(k_1+k_2). \tag{9}$$

The rational values of κ_1 and κ_2 guarantee that the phases in the two terms in Equation (7) are commensurable and the common period is the period of Φ_c, i.e.

$$T = 2\pi(k_1+k_2)\hbar/(E_1-E_2). \tag{10}$$

Unless the energy spacing between the eigenstates are commensurable, these are the only GIPQ states in the full Hilbert space.

The wave functions in (7) with nonzero κ_1 and κ_2 have no correspondence with the eigenstates. However, it is found in Ref.6) that they can be averaged to zero when are integrated over time for one period. Note that this holds only for the GIPQ solutions. Other linear combinations of the two states Ψ_1 and Ψ_2 cannot be averaged to zero because the phases in the two states are incommensurable.

For the case where the energy spacings are commensurable, more GIPQ solutions can be constructed. The SHO wave packets in examples (i) and (ii) of Section 3 are such cases. Even for these cases, the time average reduces the time-dependent wave packets to the exact eigenstates with the same energy [6].

Therefore, in the full Hilbert space the set of time-averaged GIPQ solutions is identical to the set of eigenstates.

5. Outlook

So far, the GIPQ method has been quite encouraging. The formulation is simple yet general. It does not rely on any small amplitude or adiabatic assumptions. In the examples in Section 3, the GIPQ spectra are either equal to or in good agreement with the exact spectra. Of course, one should keep in mind that the method is based on the variational principle. As it is true for all variational methods, the absolute accuracy depends decisively on the manifold in which the wave function is being varied.

The time average discussed in Section 4 is seen to be able to reduce the GIPQ exact Schrödinger solutions to the exact eigenstates. The time average of the GIPQ solutions in TDHF or other submanifolds in the Hilbert space should be studied.

References

1) P. Bonche, S.E. Koonin and J.W. Negele, Phys. Rev. C13 (1976) 1226

2) K.-K. Kan, J.J. Griffin, P.C. LIchtner and M. Dworsecka, Nucl. Phys. A332 (1979) 109

3) K.-K. Kan, Phys. Rev. C22 (1980) 2228

4) K.-K. Kan, Phys. Rev. C24 (1981) 279

5) K.-K. Kan, Phys. Rev. C24 (1981) 789

6) K.-K. Kan, J.J. Griffin, T.W. Atwater and M. Dworzecka, Phys. Rev. A27 (1983) 12

7) P.C. Lichtner, J.J. Griffin, H. Schultheid, R. Schultheis and A.B. Volkov, Phys. Lett. 88B (1979) 221; Phys. Rev. C20 (1979) 845

8) K.-K. Kan, Phys. Rev. A24 (1981) 2831

9) R. Gilmore and D.H. Feng, Phys. Lett. 76B (1978) 26

10) K.-K. Kan, P.C. Lichtner, M. Dworzecka, J.J. Griffin, Phys. Rev. C21 (1980) 1098

11) K.-K. Kan and J.J. Griffin, Phys. Rev. C15 (1977) 1126; Nucl.Phys. A301 (1978) 258; K.-K. Kan, PhD. thesis (University of Maryland, 1975)

12) J. Broeckhove, M. Buysse and P. Van Leuven, Phys. Lett. 134B (1984) 379

PERSPECTIVES OF GAUSSIAN WAVE PACKET TECHNOLOGY FOR POLYATOMIC SYSTEMS [*]

L. Lathouwers [**]

Dienst Teoretische en Wiskundige Natuurkunde

University of Antwerp, R.U.C.A.,

Groenenborgerlaan 171, 2020 Antwerp, Belgium.

Among the variety of applications of wavepackets in quantum mechanics, their use as basis functions in the molecular nuclear motion problem offers a number of promising perspectives. Building on the theory of generator coordinates as applied to molecules Heller and Blanco [1] suggested to combine angular momentum projection in combination with wave packet basis sets to treat the motion of nuclei within the Born-Oppenheimer approximation. This type of approach presents an alternative to the usual one, known as frame transformation theory, in which the kinetic energy of the nuclei is expressed in terms of the center of mass vector, angular momentum components along rotating axes and a set of $3N-6$ internal coordinates. It is important to follow up these ideas because frame transformation theory rapidly becomes cumbersome to apply as the number of nuclei increases. Indeed, there is an extensive literature on three particle systems but only a few attempts for four particle problems [2]. This is due to the fact that the transformation of the nuclear kinetic energy operator from a cartesian representation to one in generalised coordinates has to be done analytically and therefore involves tedious manipulations of the coordinate transformation in order to transform second order differential operators. Moreover, the entire procedure has to be redone for every choice of rotating frame and/ or internal variables. This makes experimenting with different choices of coordinate transformations, in order to find the physically most relevant one, very problematic. It is our purpose here to demonstrate that a wave packet representation of the motion of nuclei on a potential energy surface may be a way out of the difficulties associated with frame transformation theory. The procedure described is modelled after the use of cartesian Gaussian basis

[*] Work supported in part by the NATO Scientific Affairs Division

[**] Senior Research Associate of the N.F.W.O., Belgium.

functions in molecular electronic structure theory.

In a polyatomic system consisting of N nuclei with position vector $\bar{R}(i)$ one can associate with every nucleus a one-particle basis function

$$\chi(\bar{R}(i)|\bar{\alpha}(i),S_i) = e^{-2S_i(\bar{R}(i)-\bar{\alpha}(i))^2} . \qquad (1)$$

In the terminology of quantum chemistry one would refer to (1) as a 1s-Gaussian orbital centered at $\alpha(i)$ with exponent $2S_i$. The use of cartesian Gaussians in electronic structure calculations is mainly motivated by the analytical availability of multicenter integrals. In nuclear motion problems one has the additional advantage that the Gaussian function is intimately related to the vibrational dynamics of polyatomic systems which in zeroth order reduces to a set of uncoupled oscillators. In order to fully exploit this feature of the Gaussian function one must allow for complex centers in (1). Indeed, aside from normalisation factor we can rewrite (1) as

$$e^{-2S_i(\bar{R}(i)-Re\bar{\alpha}(i))^2 + 4iS_i \bar{R}(i).Im\bar{\alpha}(i)}$$

$$\sim e^{-2S_i(\bar{R}(i)-Re\bar{\alpha}(i))^2} \begin{array}{c} \cos \\ \sin \end{array} (4S_i\bar{R}(i).Im\bar{\alpha}(i)) . \qquad (2)$$

In these forms it is clear that through a judicious choice of the complex part of the centers $\bar{\alpha}(i)$ the nodal pattern of a vibrational state can be simulated. Also the last representation, involving trigonometric functions, shows that one can use complex centers formally but still compute all quantities with real basis functions.

From the one-particle basis (1) one can form many-particle basis functions

$$\chi(R|\alpha,S) = \prod_{i=1}^{N} \chi(\bar{R}(i)|\bar{\alpha}(i),S_i) . \qquad (3)$$

Here R and α represent all particle positions and basis function centers but S represents a single scale parameter. Indeed it is easy to show that in order to properly describe the center-of-mass motion one needs to correlate the individual scale factors to the nuclear masses such that (M_0=average nuclear mass)

$$S_i = (\frac{M_i}{M_0})^{1/2} S . \qquad (4)$$

Aside from translational symmetry one can also immediately take care of
particle statistic by symmetrisation and/or antisymmetrisation over identical
boson and/or fermion nuclei. Similarly good parity states can be constructed
by including basis functions corresponding to inverted centers.

Having set up a many particle Gaussian basis set one must compute matrix
elements with respect to the Born-Oppenheimer Hamiltonian

$$H = \sum_{i=1}^{N} - \frac{1}{2M_i} \Delta_{\bar{R}(i)} + V(\bar{R}(1),\bar{R}(2),\ldots,\bar{R}(N)) \tag{5}$$

where the potential $V(R)$ is assumed to be available from an electronic
structure calculation. It is absolutely trivial to extend the existing formula
for 1s-Gaussian orbitals to the overlap and kinetic energy integral in the
present scheme. The potential energy, however, is an extremely complex many-
body operator to be computed pointwise from program packages such a Gaussian 82[3)].
Fortunately, it is possible to construct an effective approximation scheme for
the potential energy integrals. For this purpose one can use the general
gradient expansion around a configuration $s(1),s(2),\ldots,s(N) \equiv s$

$$V(R) = \sum_{P} P! \sum_{[P]} V_{[P]}(s) \prod_{ik} (R_k(i)-s_k(i))^{P_k(i)} / P_k(i)! \tag{6}$$

where $[P] \equiv [p_x(1),p_y(2),\ldots,p_z(N)]$ is a partition of the gradient order P
and $V_{[P]}(s)$ the corresponding partial derivatives at s. These quantities are
nowadays readily available from electronic structure packages. A term by
term integration of (6) with a product of two many particle Gaussians (3)
yields an expression that factorises in both particles and cartesian directions. The result is a product of integrals of the form

$$\int_{-\infty}^{+\infty} e^{-4S(x-s)^2}(x-Res)^P \, dx = (\tfrac{i}{4})^P(\tfrac{1}{S})^{P/2} \, H_p(\sqrt{4S} \, Ims) \tag{7}$$

where the H_p are the Hermite polynomials. It is easy to estimate the convergence rate of the term by term integration of the gradient expansion (6).
Indeed, if the width of the wave packets is to be of the same order of magnitude
as the amplitude of the molecular vibrations we can put $S \cong 1/\kappa^2$ where
$\kappa = (m/M_0)^{1/4}$ is the Born-Oppenheimer parameter. This implies that, since in
most molecules $\kappa \cong 1/10$, replacing the potential energy by its gradient expansion

to order P gives the potential energy matrix elements to 10^{-P} accuracy.
At present first and second order gradients are readily available while routines
for third and fourth order derivatives are being developed. It is important
to observe that these quantities are produced at a fraction of the cost of
computing the actual potential energy. Therefore the approximate evaluation
of the potential energy matrix elements is expected to be both accurate and
cost effective.

So far we have not considered rotational invariance which in frame
transformation theory is the main obstacle to the development of a practical
computational scheme. Within the framework of the present approach one can
solve this problem by restricting the Gaussian wave packet basis (3) to those
linear combinations that properly transform under rotations. These rotational-
ly adapted basis functions are obtained via angular momentum projection

$$\chi_{MK}^{J}(R|\alpha) = P_{MK}^{J} \chi(R|\alpha) = \frac{2J+1}{8\pi^2} \int d\Omega \ D_{MK}^{J}(\Omega) \overset{*}{R}(\Omega) \chi(R|\alpha) \tag{8}$$

where $R(\Omega)$ and $D_{MK}^{J}(\Omega)$ are the rotation operators and rotation matrices depending on the Euler angles Ω. Due to the properties of the angular momentum
projection operator P_{MK}^{J} the $\chi_{MK}^{J}(R|\alpha)$ are   total angular momentum eigen-
states, i.e., eigenfunctions of J^2 and J_z with corresponding quantum numbers
J and M. Consequently, the eigenstates of (5) can always be found in the form

$$\psi_{JM}(R) = \sum_{K} \int d\alpha \ F_{K}^{J}(\alpha) \ P_{MK}^{J} \chi(R|\alpha) \quad . \tag{9}$$

Application of the variational principle to the superposition amplitudes
$F_{K}^{J}(\alpha)$ yields a set of coupled integral equations

$$\sum_{L} \int [H_{KL}^{J}(\alpha,\beta) - E^{J} \Delta_{KL}^{J}(\alpha,\beta)] F_{L}^{J}(\beta) d\beta = 0 \tag{10}$$

$$H_{KL}^{J}(\alpha,\beta) = \int d\Omega \ D_{KL}^{J}(\Omega)^{*} \langle \chi(\alpha)|HR(\Omega)|\chi(\beta)\rangle \tag{11}$$

$$\Delta_{KL}^{J}(\alpha,\beta) = \int d\Omega \ D_{KL}^{J}(\Omega)^{*} \langle \chi(\alpha)|R(\Omega)|\chi(\beta)\rangle \tag{12}$$

whose solution yields the eigenstates of (5) with total angular momentum J.
In practice the angular momentum projection scheme necessitates the calculation of matrix elements involving rotated Gaussian wave packets and the inte-

gration of these quantities with the irreducible representations of the rotation group. The first step, i.e., the calculation of $\langle\chi(\alpha)|HR(\Omega)|\chi(\beta)\rangle$ and $\langle\chi(\alpha)|R(\Omega)|\chi(\beta)\rangle$ is greatly facilitated by the following property of the many-particle basis (3)

$$R(\Omega)\ \chi(R|\alpha) = \chi(A^{-1}(\Omega)R|\alpha) = \chi(R|A(\Omega)\alpha) \tag{13}$$

where $A(\Omega)$ represents the orthogonal transformation corresponding to $R(\Omega)$. This property allows us to compute matrix elements for rotated Gaussians by simply rotating the centers of the wave packets, i.e.,

$$\langle\chi(\alpha)|HR(\Omega)|\chi(\beta)\rangle = \langle\chi(\alpha)|H|\chi(A(\Omega)\beta)\rangle \tag{14}$$

$$\langle\chi(\alpha)|R(\Omega)|\chi(\beta)\rangle = \langle\chi(\alpha)|\chi(A(\Omega)\beta)\rangle \quad . \tag{15}$$

These quantities are, however, extremely complex functions of the Euler angles Ω such that the analytical evaluation of the integrations in (11) and (12) is not feasible. One must therefore rely on numerical integration or use a technique known as approximate angular momentum projection. In the latter scheme one computes the integrals (11) and (12) perturbatively, i.e.,

$$H_{KL}^{J}(\alpha,\beta) = \sum_{N} H_{KL}^{J}(\alpha,\beta|N) \tag{16}$$

$$\Delta_{KL}^{J}(\alpha,\beta) = \sum_{N} \Delta_{KL}^{J}(\alpha,\beta|N) \quad . \tag{17}$$

It can be shown that when applied to the present problem each term in (16) and (17) is of order $(1/S)^{2N} \cong \kappa^{4N}$ and involves matrix elements of the form

$$\langle\chi(\alpha)|H\ J_x^k J_y^\ell J_z^m|\chi(\beta)\rangle \quad \text{and} \quad \langle\chi(\alpha)|J_x^k J_y^\ell J_z^m|\chi(\beta)\rangle \tag{18}$$

where J_x, J_y and J_z are total angular momentum components and $k+\ell+m \leq 2N$. The new quantities (18) can be reduced to single particle integrals which in turn can be evaluated analytically. Indeed powers of angular momentum operators

working on 1s Gaussian functions produce linear combinations of higher order
(p,d,f,...) Gaussian orbitals. The integral formula for the latter are contained
in electronic structure packages. Thus approximate angular momentum projection
is much like the approximate evaluation of potential energy matrix elements:
it is rapidly converging and feasible using existing integral subroutines.

Assuming that the integral kernels $H_{KL}^{J}(\alpha,\beta)$ and $\Delta_{KL}^{J}(\alpha,\beta)$ are available the
solution of (10) is still a formidable problem. Indeed a discretisation of
the integrals or an expansion in basis functions of the amplitudes $F_{K}^{J}(\alpha)$ will
yield secular equations of dimension $(2J+1) \times D^{3N}$ if D is the number of discretisation points or basis functions in each of the variables $\alpha_{k}(i)$.
In addition this approach has the drawback, typical of any brute force variational approach, that the quality of the results (energy levels and wave
functions) decreases with increasing excitation energy. The semi-classical
interpretation of Gaussian wave packets indicates that there might be a more
sensible way of attacking the solution of (10). Let us compare the classical
Hamilton equations of motion with the Ehrenfest theorem applied to the Gaussian
basis states (3):

$$\frac{dR_{k}(i)}{dt} = \frac{P_{k}(i)}{M_{i}} \qquad\qquad \frac{dRe\alpha_{k}(i)}{dt} = \frac{4S_{i}}{M_{i}} \text{ Im } \alpha_{k}(i)$$

$$\tag{19}$$

$$\frac{dP_{k}(i)}{dt} = -\frac{\partial V}{\partial R_{k}(i)}(R) \qquad\qquad 4S_{i}\frac{d \text{ Im } \alpha_{k}(i)}{dt} = -\frac{\partial V}{\partial R_{k}(i)}(Re\alpha) + \ldots$$

Observe that, aside from the extra terms due to expectation values of higher
order potential energy derivatives, these equations are identical if one
identifies the particle positions and momenta $\{R_{k}(i), P_{k}(i)\}$ with the wave
packet parameters $\{\text{Re } \alpha_{k}(i), 4S_{i} \text{ Im } \alpha_{k}(i)\}$. One can use this correspondence
to generate a quantum mechanical basisset associated with classical trajectories
by selecting Gaussian basis functions according to the above relationships.
This idea has been tested on one dimensional problems with promising results [5].
An interesting aspect of this scheme is that high accuracy is obtained for
levels near the energy at which the classical trajectories are run.
The accuracy fades away both for higher excitations and for lowerlying states.
It should be remarked that more general equations than (19), including a time
dependent complex scale factor S, can be derived from the so called time-
dependent variational principle. These type of approaches might help to
circumvent the high dimensionality of the secular equations and offer the

possibility of generating high quality results in predescribed spectral regions.

Clearly, the above outline of a general wave packet theory for polyatomic systems needs a lot of work both on coordination and the details of the different steps. However, it seems a feasible alternative to frame transformation theory that is more adapted to the computational methods used in theoretical chemistry.

References

1) M. Blanco and E. Heller, J. Chem. Phys. <u>78</u> (1983) 2504
2) R.J. Whitehead and W.C. Handy, J. Mol. Spectroscopy <u>59</u> (1976) 459
3) Obtainable from Prof. J. Pople, Department of Chemistry, 4400 Fifth Avenue, Pittsburgh, Pennsylvania 15213.
4) L. Lathouwers, J. Phys. <u>A18</u> (1985) 765
5) M. Davis and E. Heller, J. Chem. Phys. <u>71</u> (1979) 3383

APPLICATION OF THE TIME-DEPENDENT WAVE PACKET METHOD TO COLLISION INDUCED DISSOCIATION CALCULATIONS

C. Leforestier

Laboratoire de Chimie Théorique, Université de Paris-Sud
91405 Orsay Cedex, France

1. Introduction

Collision Induced Dissociation (CID) is important in high energy gas kinetics (e.g. explosions and flames [1], or shock tube experiments[2]) or when one of the colliding partners is only weakly bound such as Van der Waals systems[3]. It is also of fundamental importance for an understanding of basic chemical kinetics [4,5] as it constitutes the primary step of radicals formation and corresponds to the inverse process of the three body recombination. Recently, experiments have determined the binding energies of Cesium halides from the threshold for CID by rare gas atoms [6].

On the theoretical side, exact quantum 3D calculations are out of scope for the present time. Even for the simplest system A+BC → A+B+C, the exact one dimensional treatment (i.e. collinear or perpendicular collisions) has been a challenging problem until recently: Only a few of such calculations have appeared in the litterature so far [7-22], most of them being concerned with collinear non reactive systems, modelling an $H_2 + R_g$ type collision.

Three different methods have mainly been used for these calculations, namely
  i) the Time Dependent Wave Packet method [7-10]
 ii) a time independent one based on hyperspherical coordinates [16-17]
    (For a review of Close Coupling type methods, see the paper of Diestler [19]).
iii) the semi-classical (Classical S-matrix) method [20-22].

Rotational degrees of freedom (i.e. 2D or 3D calculations) are known to be important for a correct description of molecular energy transfer since the early work of Kelley and Wolfsberg[23] and Bergeron and Chapuisat [24]. This puts an important restriction on the validity of the dissociation mechanism as obtained from 1D (perpendicular or collinear) calculations. Nevertheless such quantum calculations are, for the present time, the only way to estimate the adequacy of classical mechanics to the description of dissociation processes. Also, as will be discussed in the conclusion, these collinear calculations open the way to more realistic developments (e.g. approximate 3D treatments), thanks to a new generation of (vector) computers.

In section II, we will review the time dependent wave packet method, which is now well codified since the early work of Mc Cullough and Wyatt [25]. Section III will be devoted to a comparison of classical and quantum results for reactive and non reactive collinear calculations. Finally we will discuss the possibility of extending these calculations to more realistic cases, such as approximate 3D quantum treatments.

2. The time-dependent wave packet method

The collinear A+BC dissociative collision can be treated in a straight-forward manner, using the Time Dependent Wave Packet (TDWP) method. The reason is that the dissociative continuum of the BC molecule is handled automatically within the space discretization scheme on the grid. As the basic method has already been described in detail elsewhere [25], it will be only outlined here, emphasizing the technical points and some new features which lead to a significant reduction in computation time.

2.1. INITIAL CONDITIONS

The method consists in computing the time evolution of a wave packet ψ, which represents initially an atom A impiging on a molecule BC in a given vibrational state v:

$$\psi(x,y,t = 0) = F(x,x_0).u_v(y) \tag{1}$$

where the (x,y) coordinates, which diagonalize the kinetic energy operator, can be either the usual skewed coordinates [26] for a collinear collision, or the (r,R) coordinates for a perpendicular one [22].

Most of the calculations have used a Gaussian k-distribution centered around the mean value k_0:

$$F(x,x_0) = \{\hbar/2\pi\}^{1/2} \int_{-\infty}^{+\infty} \phi(k,k_0)\exp\{-ik(x-x_0)\}dk$$

$$= \{2\pi\delta^2\}^{-1/4} \exp\{- \frac{(x-x_0)^2}{4\delta^2} - ik_0x\} \tag{2}$$

$$\phi(k,k_0) = \{2\delta^2/\pi\}^{1/2} \exp\{-2\delta^2(k-k_0)^2\} \quad . \tag{3}$$

As is well known, this initial distribution in momentum space allows to extract, from the final wavepacket, results for a whole range of collision energies $\hbar^2 k^2/2\mu_{A,BC}$ centered around the mean collision energy $\hbar^2 k_0^2/2\mu_{A,BC}$. The reliability of the results as a function of the distance $|k-k_0|$ will be discussed in the next paragraph.

Other translational distribution functions F can be defined: e.g. Mazur and Rubin [27] have used the Fourier transform of a Maxwell-Boltzmann energy distribution. Recently Raff and collaborators [28] have obtained directly rate coefficients by using Boltzmann distributions both for the translational energy and the vibrational populations.

2.2 PROPAGATION OF THE WAVEPACKET

The time evolution of the initial wavepacket (eq.1) is given in terms of the evolution equation:

$$\psi(t) = \exp\{-iH(t-t_0)/\hbar\}\psi(t_0) \quad . \tag{4}$$

To treat the space dependence of this partial differential equation, one defines a grid

$$\{x_0, x_0+\Delta x, \ldots, x_0+M.\Delta x\} \times \{y_0, y_0+\Delta y, \ldots, y_0+N.\Delta y\}$$

and requires that the solution satisfies the above equation at each node. The second derivatives at each node (x_p, y_q) are computed according to the finite difference scheme:

$$\partial^2\psi/\partial x^2\Big|_{x_p,y_q} = \Delta x^{-2}\{D_0^{(d)}\psi_{pq} + D_1^{(d)}(\psi_{p+1,q}+\psi_{p-1,q})+\ldots$$

$$+ D_d^{(d)}(\psi_{p+d,q}+\psi_{p-d,q})\} \tag{5}$$

where d is the finite difference order ($d \geq 1$) and the $D_n^{(d)}$ are the corresponding coefficients. Tests have shown that high order schemes ($d \geq 3$) can lead to a significant increase of the mesh size ($\Delta x, \Delta y$) for a same accuracy [10,29].

Recently Kosloff and Kosloff [30] have proposed an alternate way to evaluate the second derivatives by means of a pseudo spectral method: it consists
of Fourier transforming the wave function ψ into momentum space, then to
compute the second derivative by multiplication by k_i^2 and to transform back
to the spatial domain. This method should allow to use larger mesh sizes and
is particularly well suited for vectorization.

Two different implicit schemes have been shown to be particularly
efficient in order to propagate the wave packet in time:

i) the Richardson scheme [31-32] which corresponds to a 3rd order Taylor
expansion of the evolution operator (eq.(4))

$$\psi(t+\Delta t) = \psi(t-\Delta t) - 2i\Delta t/h.\underset{\sim}{H}\psi(t) + 0(\Delta t^3) \qquad (6)$$

ii) an expansion of the evolution operator in a Chebychev series [33]

$$\exp\{-iH\Delta t\} = \sum_{n=0}^{N} a_n \phi_n(-iH\Delta t) \qquad (7)$$

where the a_n are the expansion coefficients and the ϕ_n the complex Chebychev
polynomials. This scheme is specially adapted to long propagation because the
error decreases exponentially once N is large enough. The error in propagation
can thus be kept lower than the round-off error imposed by the computer or the
error arising from the spatial discretisation scheme.

2.3 FINAL ANALYSIS

When the reaction is complete, one can extract the state-to-state probabilities by projecting the final wave packet $\psi(x,y,T)$ onto the asymptotical
eigenstates. For example the inelastic probabilities $P_{v \to v'}(k)$ for the collision
energy $E = h^2 k^2/2\mu_{A,BC}$ are computed from the formula

$$P_{v \to v'}^{I}(k) = k/k_{v'}\{4\pi^2 |\phi(k,k_0)|^2\}^{-1} |\iint dx\ dy\ u_v(y) \times \exp(-ik_v x)\psi(x,y,T)|^2 \qquad (8)$$

where $\phi(k,k_0)$ is the distribution of the initial wave packet in momentum space.
A similar formula [10] holds for the reactive probabilities $P_{v \to v''}(k)$. The
dissociation probability is then computed from the relation

$$P_v^D(k) = 1 - \sum_{v'} P_{v \to v'}^I(k) - \sum_{v''} P_{v \to v''}^R(k) \quad . \tag{9}$$

In order to measure the accuracy of the calculations, several tests can be performed [9-10].

i) the first one consists in checking out if the inelastic and reaction probabilities sum up to 1 below the dissociation energy.

ii) a second test is to compare probabilities computed from different wave packets with an overlapping distribution in momentum space.

iii) one can also check if the microreversibility principle is verified by comparing the state-to-state probabilities $P_{v \to v'}(k_v)$ and $P_{v' \to v}(k_{v'})$ arising from two different calculations.

These three tests allow to define the effective range of collision energies for the final analysis.

3. Results

In this section we will discuss separately the results obtained for non reactive systems from those obtained for reactive ones, as they display very different behaviors.

3.1. NON REACTIVE SYSTEMS

The results reported [8,9,21] correspond to Rg-H_2 model systems, but where the mass of the rare gas atom Rg has been set to 1 a.u.m. Amongst these systems is included the H-H_2 one using a L.E.P.S. potential energy surface, as studied by Kulander et al. [8], because it displays no reactive scattering for the collision energies considered. All these calculations lead to the similar trends, exemplified on fig.1, namely

i) the dynamical threshold is much higher than the energetic one when the diatomic is initially in its ground vibrational state. This feature is an artefact of collinear calculations: both experimental evidence and 3D quasi-classical results for similar systems [6,34] have shown that these two thresholds usually coincide. The high threshold energies observed in these calculations decrease rapidly however with initial excitation of the diatomic.

ii) CID is strongly enhanced at low energies and tends to be inhibited at high energies. This feature has been discussed by Hunt and Child [35] using a classical S-matrix phase-space approach. Briefly the underlaying mechanism can be explained as follows: in a non-reactive $A+BC(v_0) \to A+BC(v)$ collision,

the higher the initial state v_0, the broader will be the corresponding final
distribution over vibrational states v. Henceforth the final distribution
associated to an initially excited v_0 state will hit the dissociation limit
first. When increasing the total energy, the v_0=0 state will eventually
lead to a nearly complete dissociation. At this same total energy, the
broader final distribution associated to the high initial v_0 state will still
display some components on the highest vibrational states and thus results
in a vibrational inhibition.

iii) by comparison with quasi-classical results, the quantum tails for
dissociation are larger when the diatomic is initially in a low vibrational
state. Above the threshold region, there is a reasonable agreement between
the quasi-classical and quantum results, except for the oscillations which
are not reproduced classically. Such oscillations have already appeared in
the CID study of a truncated harmonic oscillator by Johnson and Roberts [20]
and in the dissociation of a forced Morse oscillator by Hunt and Stridharan [36]
In each case the number of bumps in the dissociation curve was found to be
equal to the initial vibrational number of the oscillator.

3.2. REACTIVE SYSTEMS

To date only three reaction-dissociation calculations have been reported
by Manz and Romelt [16], Kaye and Kuppermann [17] and Leforestier [10,37].
The former two used the hyperspherical coordinates method to study model
$X-X_2$ systems, bearing respectively one and two-bound states asymptotically.
The latter one, whose results are presented on figure 2, will be ducussed
below; it corresponds to a model H-HD system, HD and H_2 bearing respectively
7 to 6 bound states.

Figure 2 enlights two distinct features from the inelastic case (see
figure 1 for a comparison) which are

i) the near coincidence of the energetic and dynamics thresholds for
dissociation. This feature has not been observed for inelastic systems.
Even on a reactive surface, when no reactive scattering occurs at the
dissociation energy, no coincidence between the two thresholds is observed [8].
The condition for the near coincidence of energetic and dynamic thresholds
is the presence of concomitant reactive scattering at the dissociation energy.
It is related to the fact that energy transfer is more efficient in reactive
collisions.

ii) smal quantum tails at the dissociation threshold but important quantum effects at higher energies. Just above threshold and up to twice the dissociation energy, there is a very good agreement between the classical and quantum results. But unlike the inelastic case, one can note large discrepancies at higher total energies. These discrepancies, which appear as sharp peaks in the classical dissociation probability curves, are due to anti-threshold effects [38,39]: these peaks result from the falling off of the inelastic probability curve. One can see however that the higher the initial vibrational state of the diatomic molecule, the better is the agreement between the classical and quantum curves.

4. Discussion

The quantum calculations reported so far help to determine the validity of a classical description of CID. In the inelastic case, one should expect quantum effects at threshold, this effect having been found more important for the (1,2,1) mass combination than for the (1,1,1) one [9]. These threshold effects can produce large errors in the calculation of thermal rate constants. The results obtained at higher energies display a vibrational inhibition effect and show there is a reasonable agreement between the classical and quantum curves. On the converse, the results obtained in the reactive case display small quantum tails in the threshold region for dissociation, but large quantum effects at higher energies due to anti-threshold effects. Such effects could persist in 3D calculations whenever the reaction cross section collapses at some energy. More calculations on reactive-dissociative systems need to be performed in order to get a clearer picture of these effects.

While only one-dimensional exact calculations are feasible for the present time, approximate 3D quantum treatments can be used. The Infinite Order Sudden Approximation [40] appears to be particularly well suited to study the dissociation of a heavy diatomic collided by a light atom: This approximation considers that the diatomic does not rotate during the collision. Within this approximation, the full 3D quantum treatment is replaced by a series of 1D calculations corresponding to

i) different values of the angle γ between the axis of the diatomic and the direction of the incident atom,

ii) different values of the relative angular momentum ℓ.
The dissociation cross section is then computed from the corresponding dissociation probabilities averaged over γ and ℓ.. Such a study is presently undertaken on the He-Ar$_2$ system.

References

1) R.D. Kern in "Comprehensive Chemical Kinetics", vol.18, edited by
C.H. Banford and C.F.H. Tipper, Elsevier, New York, 1976

2) G.W. Mc Clure a,d J.M. Peck, "Dissociation in Heavy Particle Collisions",
WIley, New York, 1972

3) M. Yamashita, T. Sano, S. Kotake and J.B. Fenn, J. Chem. Phys., 75
(1981) 5355

4) D.L. Bunker, "Theory of Elementary Gas Reaction Rates", Pergamon,
New York, 1966

5) V.N. Kondratiev and E.E. Nikitin, "Gas-Phase Reactions", Springer-Verlag
Berlin, Heidelberg 1981

6) E.K. Parks, S. Wexler, J. Phys. Chem., 88 (1984) 4492

7) L.W. Ford, D.J. Diestler and A.F. Wagner, J. Chem. Phys. 63 (1975) 2019

8) K.C. Kulander, J. Chem. Phys., 69 (1978) 5064;
J.C. Gray, G.A. Fraser, D.G. Truhlar, K.C. Kulander, J. Chem. Phys.,
73 (1980) 5726

9) C. Leforestier, G. Bergeron and P. Hiberty, Chem. Phys. Lett. 84 (1981) 385;
G. Bergeron, P. Hiberty and C. Leforestier, Chem. Phys. 93 (1985) 253

10) C. Leforestier, Chem. Phys. 87 (1984) 241

11) G. Wolken, J. Chem. Phys. 63 (1975) 528

12) E.W. Knapp, D.J. Diestler and Y.W. Liu, Chem. Phys. Lett. 49 (1977) 379

13) E.W. Knapp and D.J. Diestler, J. Chem. Phys. 67 (1977) 4969

14) L.H. Beard and D.A. Micha, J. Chem. Phys. 73 (1980) 1193

15) G.D. Barg and A. Askar, Chem. Phys. Lett. 76 (1980) 609

16) J. Manz and J. Römelt, Chem. Phys. Lett. 77 (1981) 172

17) J.A. Kaye and A. Kuppermann, Chem. Phys. Lett. 78 (1981) 546

18) M.I. Haftel and T.K. Lim, Chem. Phys. Lett. 89 (1982) 31

19) D.J. Diestler in :"Atom-molecule collision theory: a guid for the experimen-
talist", ed. R.B. Bernstein, Plenum Press, New York, 1979

20) L.L. Johnson and R.E. Roberts, Chem. Phys. Lett. 7 (1970) 480

21) I. Rusinek and R.E. Roberts, J. Chem. Phys. 65 (1976) 872;
ibid. 68 (1978) 1147

22) I. Rusinek, J. Chem. Phys. 72 (1980 4518

23) J.D. Kelley and M. Wolfsberg, J. Chem. Phys. 53 (1970) 2967

24) G. Bergeron and X. Chapuisat, Chem. Phys. Lett. 11 (1971) 334

25) E.A. Mc Cullough and R.E. Wyatt, J. Chem. Phys., $\underline{54}$ (1971) 3678

26) P.J. Kuntz in :"Modern Theoretical Chemistry", vol. 2, ed. W.H. Miller, Plenum Press, New York, 1976

27) J. Mazur and R.J. Rubin, J. Chem. Phys. $\underline{31}$ (1959) 1395

28) P.M. Agrawal and L.M. Raff, J. Chem. Phys. $\underline{74}$ (1981) 5076
    P.M. Agrawal, N.C. Agrawal, R. Visvanathan and L.M. Raff,
    J. Chem. Phys. $\underline{80}$ (1984) 760

29) C. Leforestier, unpublished results.

30) D. Kosloff and R. Kosloff, J. Comp. Phys. $\underline{52}$ (1983) 35

31) D.M. Young and R.T. Gregory, "A survey of numerical mathematics", Addison-Wesley, Reading 1973

32) A. Askar and A.S. Cakmak, J. Chem. Phys $\underline{68}$ (1978) 2794
    R.J. Rubin, J. Chem. Phys. $\underline{70}$ (1979) 4811

33) H. Tal-Ezer and R. Kosloff, J. Chem. Phys. $\underline{81}$ (1984) 3967

34) D.J. Malcolm-Lawes, J. Chem. Soc. Faraday Trans. II $\underline{71}$ (1975) 1183

35) P.M. Hunt and M.S. Child, J. Phys. Chem. $\underline{86}$ (1982) 1116

36) P.M. Hunt and S. Sridharan, J. Chem. Phys. $\underline{77}$ (1982) 4022

37) C. Leforestier, to be published

38) S.C. Leasure and J.M. Bowman, Chem. Phys. Lett. $\underline{39}$ (1976) 462

39) G. Bergeron and C. Leforestier, Chem. Phys. Lett. $\underline{71}$ (1980) 519

40) G.A. Parker and R.T. Pack, J. Chem. Phys. $\underline{68}$ (1978) 1585

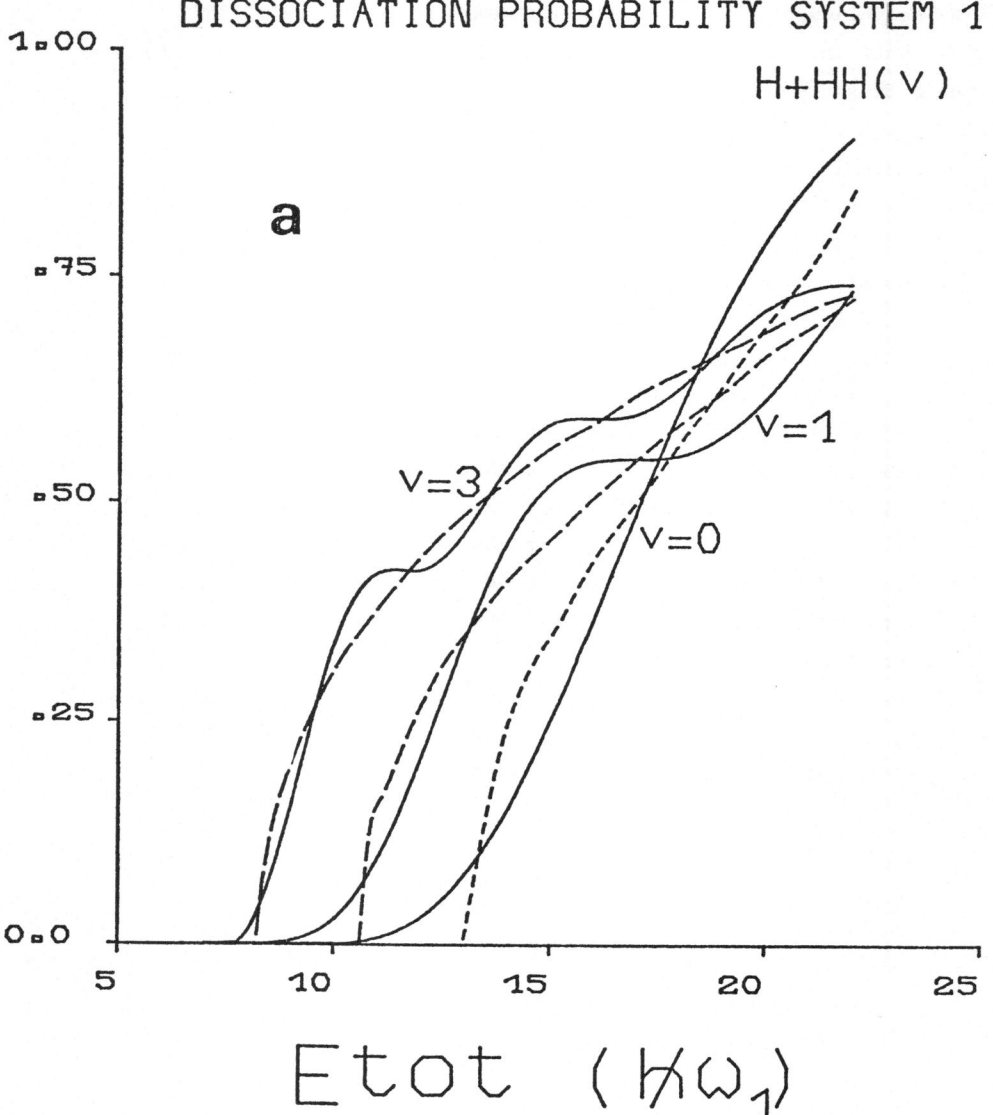

Figure 1: Comparison of quantum (——) and classical (- -) dissociation probabilities for a model H+HH(v) → H+H+H inelastic system as a function
of the total energy. The total energy is expressed in units of the
fundamental frequency ω of the model H_2 molecule.

DISSOCIATION PROBABILITY

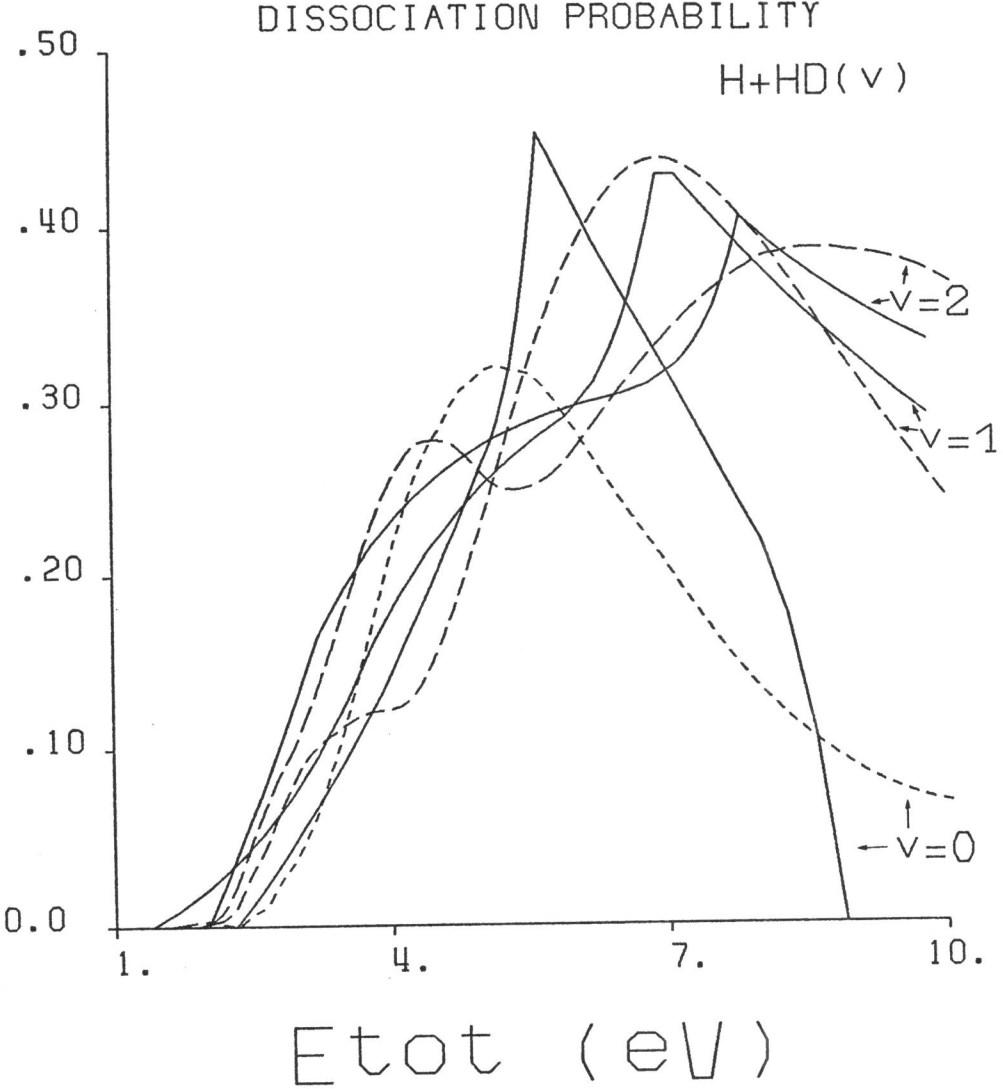

H+HD(v)

Figure 2: Comparison of quantum (- -) and classical (——) dissociation probabilities for a model H+HD(v) → H+H+D reactive system as a function of the total energy.

CREATION AND TIME EVOLUTION OF THE COHERENT ROTATIONAL STATES

N. Mankoc-Borstnik

Department of Physics, E. Kardelj University

P.O. Box 543, Jadranska 19, 61001 Ljubljana, Yugoslavia

1. Introduction

The aim of this talk is to review the properties of the coherent rotational states (CRS) connected with the peculiar time dependency. Due to the fact that the energies of the members of the ideal rotational band obey the rule $\hbar\omega J(J+1)$, the CRS express a periodic behaviour in time. Coherent superpositions of the rotational states of nuclei or molecules can represent nuclei or molecules partly or completely oriented in space. Since in the time evolution the coherence disappear and recovers again periodically, also the orientation in space disappears in time and appears again peridically. Accordingly the properties of CRS change in time and so do their interaction with the electromagnetic field or with other particles. These properties can manifest in radiation in bursts [1] in the case of nuclei and in periodic change of refractive index [2] in gases. Since the nuclei and molecules obey the rule $\hbar\omega J(J+1)$ only approximately, their properties change quasi periodically in time, rather than periodically.

2. Definition of the coherent rotational states

Using the method of boson operator [3], one first defines the vacuum state $|\psi_0\rangle$ and the two states with the angular momentum 1/2 and the third projection $\pm 1/2|x_{\pm}\rangle$ generated by the two operators $\hat{x}_{\pm}$. The operators $J_{\pm}$ lower the angular momentum of a state for 1/2 and the third projection for $\pm 1/2$.

$$\hat{x}_{\pm}|\psi_0\rangle = |x_{\pm}\rangle$$

$$\hat{J}_{\pm}|\psi_0\rangle = 0 \quad . \tag{1}$$

Operators $\hat{\chi}_{\pm}$, $\hat{\delta}_{\pm}$ fulfill the bosonic commutational relations:

$$[\hat{J}_i, \hat{\chi}_j] = \delta_{ij} \quad , \quad [\hat{\chi}_i, \hat{\chi}_j] = [\hat{\delta}_i, \hat{\delta}_j] = 0 \quad .$$

The angular momentum operator can be expressed as a bilinear forms of bosonic operators $\hat{\chi}_{\pm}$ and $\hat{\delta}_{\pm}$:

$$\hat{J}_x = \frac{1}{2} (\hat{\chi}_-\hat{J}_+ + \hat{\chi}_+\hat{J}_-)$$

$$\hat{J}_y = \frac{1}{2} (\hat{\chi}_-\hat{J}_+ - \hat{\chi}_+\hat{J}_-)$$

$$\hat{J}_z = \frac{1}{2} (\hat{\chi}_+\hat{J}_+ - \hat{\chi}_-\hat{J}_-) = \frac{1}{2} (\hat{N}_+ - \hat{N}_-) \tag{2}$$

$$\hat{J}_{\pm}^2 = \hat{\chi}_{\pm}\hat{J}_{\mp}$$

$$\hat{J}^2 = \frac{1}{2} \hat{N}(\frac{1}{2}\hat{N}+1) \quad .$$

Here $\hat{N} = \hat{N}_+ + \hat{N}_-$, and $\hat{N}$ is the operator measuring the total number of one half steps in the angular momentum expectation values. The CRS are then defined (similar to the case of the coherent vibrational states) as simultaneous eigenstates of the destruction operators J_+ and J_-

$$\hat{J}_{\pm}|\psi_\beta\rangle = \beta_{\pm}|\psi_\beta\rangle \quad , \quad \langle\psi_\beta|\psi_\beta\rangle = 1 \quad .$$

One obtains:

$$|\psi_\beta\rangle = e^{-1/2(|\beta_+|^2+|\beta_-|^2)} e^{\beta_+\hat{\chi}_+ + \beta_-\hat{\chi}_-} |\psi_0\rangle .$$

$$= e^{-1/2(|\beta_+|^2+|\beta_-|^2)} \sum_{IM} \frac{\beta_+^{I+M} \beta_-^{I-M}}{\sqrt{(I+M)!(I-M)!}} |\phi_{IM}\rangle, \tag{3}$$

with

$$|\phi_{IM}\rangle = \frac{(\hat{\chi}_+)^{I+M} (\hat{\chi}_-)^{I-M}}{\sqrt{(I+M)!(I-M)!}} |\psi_0\rangle,$$

and

$$\frac{1}{\pi^2} \int d^2\beta_+ d^2\beta_- \ |\psi_\beta\rangle\langle\psi_\beta| = I \quad . \tag{4}$$

The CRS have the following characteristic properties:

i) The absolute values of the amplitudes $c_{IM} = e^{-1/2(|\beta_+|^2 + |\beta_-|^2)}$.

$$\frac{\beta_+^{I+M} \beta_-^{I-M}}{\sqrt{(I+M)!(I-M)!}}$$

are peaked around the mean values of the angular momenta:

$$\langle\psi_\beta|\hat{J}_z|\psi_\beta\rangle = \frac{1}{2}(|\beta_+|^2 - |\beta_-|^2) \text{ and } \langle\psi_\beta|\hat{J}^2|\psi_\beta\rangle = \frac{1}{4}(|\beta_+|^2 + |\beta_-|^2)(|\beta_+|^2 + |\beta_-|^2 + 2).$$

$$\tag{5}$$

ii) The phases φ_{IM}^β of the amplitudes defined by:

$$c_{IM}^\beta = |c_{IM}^\beta| e^{-i\varphi_{IM}^\beta}$$

are exactly equidistant for fixed I (or M). As a consequence the CRS have large quadrupole moment of inertia $\beta_\pm$.

iii) If the considered vacuum state characterizes rigid rotor, all the states $|\phi_{IM}\rangle$ appearing in the CRS $|\psi_\beta\rangle$ are members of the same rotational band with rotational energies $\hbar\omega J(J+1)$. The CRS describe then a rotor completely oriented in space.

iv) The CRS have peculiar time evolution properties:

$$|\psi_\beta(t)\rangle = e^{-i\hat{H}_0 t} |\psi_\beta(0)\rangle = e^{-iE0t} \sum_{J=0}^{\infty} \sum_{M=-J}^{J} |c_{JM}^\beta| e^{-i\varphi_{JM}^\beta} e^{-i\omega J(J+1)t} |\phi_{JM}\rangle \tag{6}$$

with $\hat{H}_0|\phi_{JM}\rangle = (E_0 + \omega J(J+1))|\phi_{JM}\rangle$ and E_0 is the energy of the vacuum state. The phases $(-i\varphi_{JM}^\beta + i\omega J(J+1)t)$ periodically recover their initial values. The orientation of a rigid rotor disappears in time and recovers back periodically. The expectation values of the operators also have the periodic time behaviour

$$\langle\psi_\beta(t)|\hat{Q}|\psi_\beta(t)\rangle = e^{-(|\beta_+|^2+|\beta_-|^2)} \sum_{JM} \frac{\beta_+^{*\,J+M}\,\beta_-^{*\,J-M}\,\beta_+^{L+N}\,\beta_-^{L-N}}{\sqrt{(J+M)!(J-M)!(L+N)!(L-N)!}}$$

$$\cdot\, e^{i\omega t[J(J+1)-L(L+1)]}\cdot\langle\phi_{JM}|\hat{Q}|\phi_{LN}\rangle$$

(7)

In fig. 1 the time dependence of the quadrupole operator is presented for $\beta_+ \equiv \beta_-$ and $\langle\psi_\beta|\hat{J}^2|\psi_\beta\rangle = 110$

$\langle\psi_\beta(t)|\hat{Q}_0^2|\psi_\beta(t)\rangle$

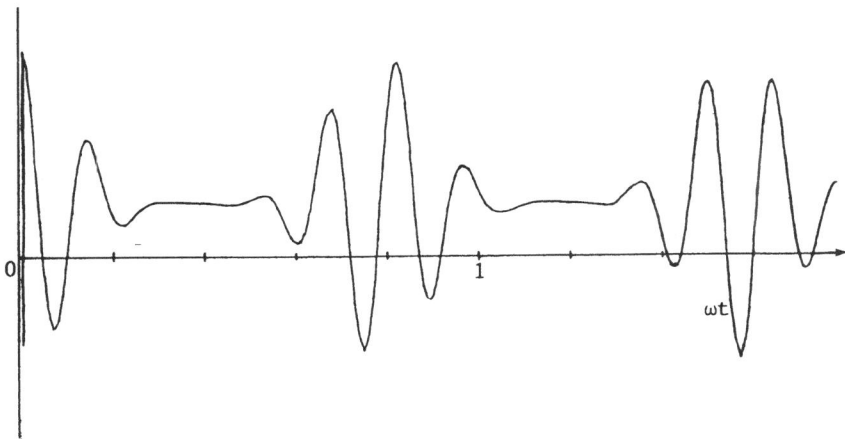

Fig. 1

3. The creation of the coherent rotational states

It turns out that under special conditions either molecules or nuclei can be excited into states which are similar to the CRS. This can happen when two nuclei with the relative kinetic energy below the Coulomb barrier scatter; in the molecular case short laser pulses can produce similar effect.

The nucleus is before scattering in a pure state:

$$\hat{\rho}_0 = |\phi_{JM}\rangle\langle\phi_{JM}| \tag{8}$$

while the molecular density operator:

$$\rho_0 = \frac{e^{-\beta\hat{H}_0}}{\text{Tr } e^{-\beta\hat{H}_0}} = \sum P_{JM}|\phi_{JM}\rangle\langle\phi_{JM}| \tag{9}$$

is determined by the temperature T of the gas.

Let $\hat{H}'(t)$ determine the interaction between the two nuclei or between the laser pulse and the molecule at $t > t_0$.

If the system is at $t < t_0$ in an eigenstate of $\hat{H}_0$:

$$\hat{H}_0|\phi_{JM}\rangle = E_J|\phi_{JM}\rangle, \tag{10}$$

the eigenfunctions of:

$$\hat{H} = \hat{H}_0 + \hat{H}' \tag{11}$$

are then:

$$|\psi^{(JM)}(t)\rangle = e^{-i\hat{H}_0(t-t_0)}\hat{U}(t_1 t_0)|\phi_{JM}(t_0)\rangle, \tag{12}$$

with

$$\hat{U}(t,t_0) = e^{i\hat{R}}, \tag{12a}$$

$$R = -\frac{1}{\hbar}\int_{t_0}^{t}\hat{H}_I'(t')dt' + \frac{i}{2\hbar^2}\int_{t_0}^{t}dt'\int_{t_0}^{t'}dt''\,[\hat{H}_I'(t'),\hat{H}_I'(t'')]$$

$$+ \text{ terms with higher order commutators.} \tag{12b}$$

Here

$$\hat{H}_I = e^{i\hat{H}_0(t-t_0)} \hat{H}' \, e^{-i\hat{H}_0(t-t_0)} .$$

If the duration τ of the pulse is short enough, so that $\dfrac{\Delta E_j \cdot \tau}{\hbar} \ll 1$, where ΔE_j is the difference between the highest excited state and the ground state, a sudden approximation can be used:

$$\hat{U}(t,t_0) = e^{-\frac{i}{\hbar} \int_{t_0}^{t} \hat{H}'(t')dt'} \quad , \quad \hat{H}_I \sim \hat{H}' . \tag{13}$$

The density matrix operator of the molecule:

$$\hat{\rho}(t) = \sum_{JM} |\psi^{(JM)}(t)\rangle\langle\psi^{(JM)}(t)| P_{JM}, t > t_0 \tag{14}$$

is determined by the time evolution of each state $|\phi_{JM}\rangle$. The statistics enter through the coefficients p_{JM} which are determined at $t=t_0$. It should be pointed out that in the sudden approximation the term $\int_{t_0}^{t} dt'\hat{H}(t')$ is taken into account in all orders, so that a multistep process is taking place (see fig. 2).

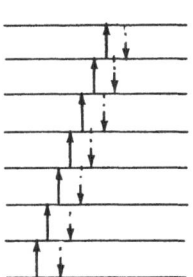

<u>Fig. 2:</u> The number of transitions is limited with the time τ of the duration of $\hat{H}'$, since each higher excited state has to wait until the previous one is populated.

4. The nuclear case

We shall study the collision between the nuclei ^{238}U and ^{40}Ca (Since ^{40}Ca has a double closed shell it will not be excited). We treat the case when the nuclear force play no role, since the kinetic energy of the relative motion is lower than the Coulomb barrier. We choose:

i) $v/c \sim 0.08$, $E \sim 100$ MeV, $\dfrac{\Delta E}{E} \sim 0.01$.

ii) The impact parameter $b \sim 25$fm, $\dfrac{\lambda [\equiv \frac{\hbar}{mn}]}{b} \sim 4.10^{-4} \ll 1$,

iii) $\dfrac{\Delta L}{L} \sim \dfrac{40}{mvb} \sim \dfrac{1}{10} < 1$.

In this case the relative motion of the two nuclei can be described by the classical trajectory.

Since $\frac{\Delta E_I \cdot \tau}{\hbar} \sim \frac{1}{20}$, the sudden approximation can be used. The interaction $\hat{H}'(t)$ is then the multipole electromagnetic excitation of the nucleus ^{238}U due to ^{40}Ca and $\hat{H}_0$ is the free Hamiltonian of the ^{238}U.

The coefficients $C_{J'M'}(t) = \langle \phi_{J'M'} | \psi^{(JM)}(t) \rangle$ have in the sudden approximation rather simple form:

$$C_{J'M'}(t) = \langle \phi_{J'M'}(t_0) | e^{-\frac{i}{\hbar} \int_{t_0}^{t} O\hat{H}'\hat{H}'(t')} | \phi_{JM}(t_0) \rangle =$$

$$= \sum_{J''M''} \langle \phi_{J'M'} | \hat{\tilde{U}} | \phi_{J''M''} \rangle e^{i\lambda_{J''M''}} \langle \phi_{J''M''} | \hat{\tilde{U}}^+ | \phi_{JM} \rangle . \tag{15}$$

Here the operator $\hat{\tilde{U}}$ is chosen to diagonalize the metrics:

$$\langle \phi_{JM} | \hat{\tilde{U}}^+ [-\frac{1}{\hbar} \int_{t_0}^{t} dt' \hat{H}'(t')] \hat{\tilde{U}} | \phi_{J'M'} \rangle = \lambda_{JM} \delta_{JJ'} \delta_{MM'} . \tag{16}$$

The results, copied from ref. [3] are presented in fig. 3.

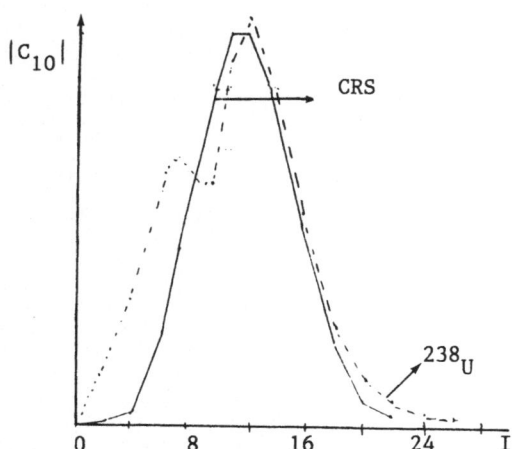

Fig. 3: Full line represents the coefficients $|c_{10}|$ of the CRS while dotted line represents ^{238}U. Only even I are considered.

5. The molecular case

The molecules of the gas CS_2 are at $t=t_0$ in the statistical equilibrium determined by the temperature $T_0 (\sim 300K)$. The coefficients P_{JM} are then: $P_{JM} = e^{-\beta EJ} / (\sum_j (2J+1)e^{-\beta EJ})$. The average populated state is the state with $J_0 \sim 40$ ($\hbar \omega J(J+1) \sim kT_0$). Many molecules can be excited into the coherent superposition of rotational states, if the pulse is spread enough. Since $\frac{\Delta EJ \cdot \tau}{\hbar} \sim 0.1$, the sudden approximation can be used.

The interaction Hamiltonian $\hat{H}'$ represents the interaction between the electromagnetic field of the laser pulse and the molecule:

$$\hat{H}'(t) = (- \sum_1 \frac{e}{2m_i} (\hat{\vec{A}}(\vec{r}_i)\hat{\vec{p}}_i + \hat{\vec{p}}_i \hat{\vec{A}}(\vec{r}_i))) \Theta(t,t_0,\tau), \qquad (17)$$

where

$$\hat{\vec{A}}(\vec{r}) = \frac{1}{\sqrt{v}} \sum_{\ell\sigma} \vec{\varepsilon}_{\ell\sigma} \frac{1}{2k_\ell} [\hat{a}^+_{\ell\sigma} e^{ik_\ell i \vec{r}} + \hat{a}_{\ell\sigma} e^{-ik_\ell i \vec{r}}]$$

with

$$\Theta(t,t_0,\tau) = \{ \begin{matrix} 1 & \text{if } t_0 < t \leq t_0 + \tau \\ 0, & \text{otherwise} \end{matrix} \quad .$$

If the magnetic dipole and electric quadrupole transitions between the molecular states are negligible in comparison with electric dipole transitions, the transitions between the two rotational states of the same band are two step Roman transitions. The coefficients c_{JM} have then a slightly different shape than in the nuclear case:

$$c_{J'M'}(t) = \sum_{J''M''} \langle\phi_{J'M'}|\hat{\tilde{U}}|\phi_{J''M''}\rangle (e^{i\lambda_{J''M''}} + e^{-i\lambda_{J''M''}}) \langle\phi_{J''M''}|\tilde{U}_+|\phi_{JM}\rangle \qquad (18)$$

where $\hat{\tilde{U}}$ now diagonalizes the matrix:

$$\langle\phi_{JM}|\hat{\tilde{U}}^+(- \frac{1}{\hbar} \int_{t_0}^t dt' H'(t'))^2 U|\phi_{J'M'}\rangle = \lambda_{JM} \delta_{JJ'} \delta_{MM'} \qquad . \qquad (19)$$

The density matrix operator has a form

$$\hat{\rho}(t) = \sum_{\substack{JM \\ J'M'}} P_{JM} c^{*(JM)}_{J'M'} |\phi_{J'M'}\rangle\langle\phi_{J''M''}| c^{(JM)}_{J''M''} \qquad (20)$$

Since the coherent states, generated on different initial states $|\phi_{JM}\rangle$ are all expected to have similar coefficients, the statistical nature of the initial density matrix should not spoil the effect. One expects the similar behaviour of the expectation values of operators then in nuclear case.

6. The time evolution of the properties of the coherent rotational states

γ-decay of the CRS of the nucleus.

We study the γ-decay of the nucleus which is at $t=t_0+\tau=0$ in the state $|\psi^{(JM)}(t)\rangle$ or $|\psi_\beta(t)\rangle$. The Hamiltonian of the system can be written in the form:

$$\hat{H} = \hat{H}_0^{nucl} + \hat{H}_0^{elm} + \hat{H}_{int} \quad . \tag{21}$$

Here $\hat{H}_0^{nucl}$ and $\hat{H}_0^{elm}$ describe the free nucleus and the free electromagnetic field, while $\hat{H}_{int}$ describes the interaction causing the γ decay of the nucleus. If $|\psi(t)\rangle$ describes the system $\hat{H}$ at $t>0$, the radiation power can be written in the form:

$$\frac{d}{dt} \langle \psi(t)| \int \hat{\mathcal{H}}dV |\psi(t)\rangle \cong - \frac{d}{dt} \langle \psi(t)| \int \mathcal{H}_0^{elm} \, dV |\psi(t)\rangle \tag{22}$$
$$\text{inside } R_0$$

and

$$\int \hat{\mathcal{H}}_0^{elm} \, dV = \sum_{\ell\sigma} k_\ell \, \hat{a}_{\ell\sigma}^+ \hat{a}_{\ell\sigma} \tag{22a}$$

If the wave function $|\psi(t)\rangle$ is written as a Taylor expansion of different number of photons:

$$|\psi(t)\rangle = \sum_\beta C_\beta(t)|\psi_\beta(t)\rangle + \sum_{\beta\ell\sigma} C_{\beta\ell\sigma}(t)|\psi_{\beta\ell\sigma}(t)\rangle + \ldots \tag{23}$$

with

$$|\psi_{\beta\ell\sigma}\rangle = |\ell\sigma\rangle|\psi_\beta\rangle.$$

and the Schrödinger equation is solved in the first order perturbation theory, the radiation power of the nucleus has the form:

$$P(t) = \frac{d}{dt} \left. \psi(t) \right| \sum_{\ell\sigma} k_\ell \hat{a}^+_{\ell\sigma} \hat{a}_{\ell\sigma} \left| \psi(t) \right\rangle = \frac{d}{dt} \sum_{\ell\sigma\beta} k_\ell |C_{\beta\ell\sigma}|^2 \qquad (24)$$

with

$$C_{\beta\ell\sigma}(t) \cong -i \int_0^t dt' \langle \psi_{\beta\ell\sigma}(t') | \hat{H}_{int} | \psi_\beta(t') \rangle,$$

and

$$\hat{H}_{int} \propto \sum_{\ell\sigma} \sqrt{\frac{k_\ell}{2}} \{ \hat{a}_{\ell\sigma} [\hat{Q}_{\ell\sigma}, \hat{H}_0] + h.c \}$$

$$\hat{Q}_{\ell\sigma} = \frac{3e}{k\cdot\ell} (\vec{k}_\ell \cdot \vec{r})(\vec{\epsilon}_{\ell\sigma} \cdot \vec{r}) \quad . \qquad (25)$$

After some algebra it follows:

$$P(t) = \frac{1}{36\pi} \sum_{IJKL} \{\Delta^5_{IJ} \Delta_{LK} \Theta(\Delta_{IJ}) + \Delta_{IJ} \Delta^5_{LK} \Theta(\Delta_{LK})\}$$

$$\overline{Q_{IJ} Q^*_{LK}} \; \frac{\sin(\Delta_{IJ} - \Delta_{LK})t/2\hbar}{\Delta_{IJ} - \Delta_{LK}} \cdot e^{i(\Delta_{LK} - \Delta_{IJ})t/2\hbar} \qquad (26)$$

where $\Theta(x)$ is 1 for $x > 0$ and 0 otherwise and $\Delta_{IJ} = E_J - E_I$.

Here $Q_{IJ} = |Q_{IJ}| e^{i(\varphi^\beta_{I0} - \varphi^\beta_{J0})}$, $Q_{IJ} = C^*_{I0} C_{J0} \langle \phi_{J0} | \hat{Q} | \phi_{I0} \rangle$, with M chosen to be 0 only and $\overline{QQ^*}$ means the average over all directions $\vec{k}/k$ and polarizations of photons.

We present in fig. 4a the radiation power of the ideal CRS and in fig. 4b the radiation power of ^{238}U.

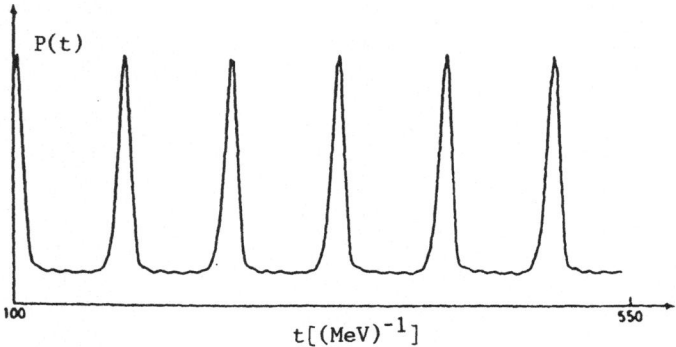

Fig. 4a: γ-decay power for a nucleus with an ideal rotational band. The constant ω is taken to be 0.02 MeV.

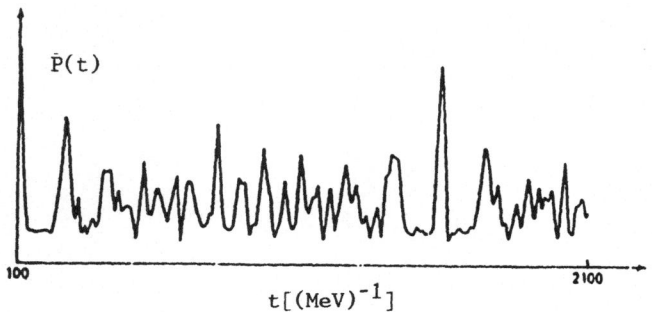

Fig. 4b: α-decay power for ^{238}U.

7. Measurements of the refractive index of the molecule in the crs

One can measure the degree of the coherence of the moleculear rotational states as a function of time by measuring the change of the polarisation of the electromagnetic pulse, travelling through the gas:

$$\mathrm{Tr}|\hat{\rho}_1(t) \sum_{\ell} |\alpha_{k\ell\sigma_1}(t)\rangle\langle\alpha_{k_\ell\sigma_1}(t)| \qquad . \tag{27}$$

Here σ_1 is pependicular to the initial polarization σ_0 of the electromagnetic pulse, and the density operator $\hat{\rho}_1(t)$ describes the testing pulse in the coherent state $|\alpha' k_0\sigma_0\rangle$ and the molecule after the interaction with the testing pulse:

$$\hat{\rho}_1(t) = \hat{U}_0(t,t_{01})\hat{U}_1(t,t_{01})\hat{\rho}(t_{01})\hat{U}_1^+(t,t_{01})\hat{U}_0^+(t,t_{01}), \tag{28}$$

where $\hat{U}_0(t,t_{01}) = e^{-i\hat{H}_0(t-t_{01})}$, with $\hat{H}_0$ being the Hamiltonian of the free (testing) electromagnetic pulse and free molecule, $\hat{U}_1(t,t_{01})$ is determined by the interaction between the testing pulse and the molecule and $\hat{\rho}(t_{01})$ is defined as follows:

$$\hat{\rho}(t_{01}) = \sum_{JM} P_{JM}|\psi^{(JM)}(t_{01})\rangle|\alpha_{k_0\sigma_0}(t_{01})\rangle\langle\alpha_{k_0\sigma_0}(t_{01})|\langle\phi^{(JM)}(t_{01})| \qquad . \tag{29}$$

Fig. 5 shows the schematic of the experimental arrangement used to detect the time dependence of the degree of coherence of the CS_2 gas at 300K. At $t=t_0$ a strong picosecond pulse of $\lambda=1.06$ μm travels through the CS_2 gas. The part of the same pulse is delayed for Δt, the polarization is changed to the perpendicular one and the higher harmonics is extracted. The weak delayed pulse travels through the gas at $t=t_0+\Delta t$ and the change of the polarization of the delayed pulse is measured. The result is presented in fig. 6. It shows the periodic nature of the coherent superposition of rotational molecular states.

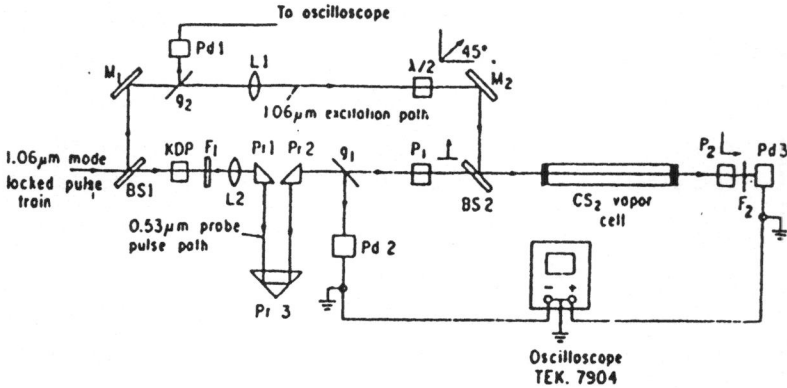

Fig.5: Schematic of the experimental arrangement used to detect the quantum beat. BS1 and BS2 are beam splitters. M_1 and M_2 are mirrors. The potassium dihydrogen phosphate (KDP) crystal was used to generate the second harmonic of the 1.06-μm pulses. F_1 and F_2 filter out residual 1.06-μm pulses. F_1 and F_2 filter out residual 1.06-μm radiation. g_1 and g_2 represent glass plates and Pd1, Pd2 and Pd3 the photodetectors. Prisms Pr1, Pr2 and Pr3 formed the adjustable delay line. P_1 and P_2 are polarizers oriented as indicated and $\lambda/2$ represents a half-wave plate. L_1 and L_2 represent 1.5-m-focal-length lenses. The CS_2 vapor cell was 1 m in length.

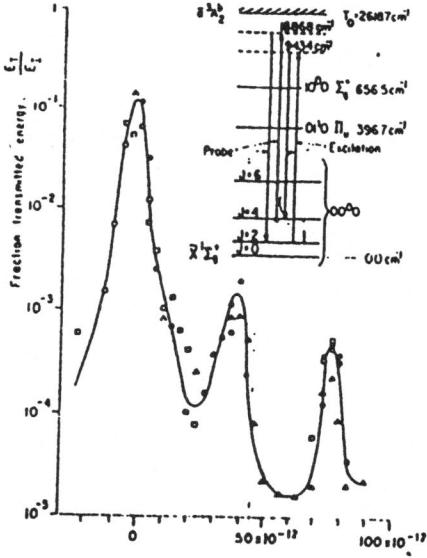

Fig. 6: The measured time dependence of the fraction of energy transmitted through the analyzing polarizer. The time origin was taken to be the observed peak in the transmitted energy. Also shown is the appropriate portion of the term diagram for the CS_2 molecule. One component of the overall Raman-mixing process is indicated (for $\Omega_1 = \Omega_2$). The dashed lines indicate virtual states lying below the excited electronic states. [See G. Hertzberg, Ref.1, p.276, and Electronic Spectra of Polyatomic Molecules (Van Nostrand-Reinhold, New York, 1968),p.801.]

8. Conclusions

Ideal coherent rotational states exhibit in time evolution the periodic structure. Nuclei and molecules with well expressed rotational spectra can be excited in multistep processes into coherent superpositions of rotational states with properties very similar to the properties of the CRS.
It was shown that such an event can occur for nuclei in heavy ion collisions below the Coulomb barrier and for molecules in the interaction with a short laser pulse. While the period of nuclei is $\sim 10^{-19}$s and therefore unmeasurable, molecular periods are $\sim 10^{-11}$s and can therefore be measured.

9. References

1) N. Mankoc-Borstnik, M . Rosina, L. Fonda,
   Nuovo Cimento 57 (1979) 440; 56A (1980) 229; 63A (1981) 483;
   58A (1980) 105.
   A review article to be published.
2) C.H. Lin, J.P. Heritage, T.K. Gustafson, Appl. Phys. Lett. 19 (1971) 397;
   Phys. Rev. Lett. 34 (1975) 1299; Phys. Rev. A13 (1976) 813.
3) P.W. Atkins, C.J. Dobson, Proc. Roy. Soc. 321A (1971) 321
4) K. Alder, A. Winther, Electromagnetic Excitation (Amsterdam, 1975).

DYNAMICS OF WAVE PACKETS AND STRONG NONADIABATIC EFFECTS

H.-D. Meyer and H. Köppel

Theoretische Chemie, Physikalisch-Chemisches Institut

Universität Heidelberg, D-6900 Heidelberg, D.B.R.

Internal conversion processes provide an important class of non-radiative transitions in molecules [1]. They consist of the decay of electronically excited states, which is caused by the nuclear motion and thus, by definition, constitute a non-adiabatic effect. The non-adiabaticity becomes very strong in the vicinity of a conical intersection. It then causes an ultrafast (i.e. sub pico-second) decay of an initial state, prepared e.g. by photo-ionizing the neutral molecule. The main question we want to answer in this talk is how the decay of a state can be described theoretically in this strong coupling case.

The conventional description of a decay of a state makes use of the correlation function [2]

$$c(t) = \langle \psi(0) | \psi(t) \rangle \quad . \tag{1}$$

We show that the correlation function is not the appropriate measure of the decay if one assumes the strong coupling limit. More appropriate measures are the occupation probabilities [3-5]

$$P_2(t) = \langle \psi(t) | P_2 | \psi(t) \rangle \tag{2}$$

and

$$P_2^{(ad)}(t) = \langle \psi(t) | P_2^{(ad)} | \psi(t) \rangle \tag{3}$$

where P_2 and $P_2^{(ad)}$ denote the projector onto the second diabatic and adiabatic electronic sub-space, respectively. These two measures, however, suffer from a certain degree of arbitrariness; one does not know which one to prefer over the other. To circumvent this problem we consider an <u>observable</u> quantity

such as the photon emission rate I(t). The prepared state (initally a gaussian
wave packet on the second diabatic potential energy surface) radiates as long
as it has not decayed. This yields an unambiguous definition of the decay.
We show [5,6] that the photon emission rate can be written as an expectation
value

$$I(t) = \langle\psi(t)|\Gamma|\psi(t)\rangle \qquad (4)$$

where the radiative damping Γ is determined by an effective Hamiltonian
formalism $H_{eff} = H - i\Gamma/2$. The effective Hamiltonian is derived via a
projection operator method from the full Hamiltonian which accounts for both
the molecule and the radiation field. The fluorescence intensity I(t) shows
an ultrafast (i.e. sub pico-second) decay and a very low long time average.
These results provide an explanation why the ethylene cation $C_2H_4^+$ as well as
most others radical cations studied to date show no detectable fluorescence
after the excitation (e.g. ionisation). A classical model [4,5] based on the
classical analog for electronic degrees of freedom [7] was also applied to the
decay problem. There is an excellent agreement between the classically and
quantally computed occupation probabilities $P_2(t)$. For the fluorescence
intensity I(t) the agreement is only qualitatively. The classical calculations serve as a useful tool to study the qualitative dependence of I(t) on
various parameters, in particular on the number of vibrational modes
(For numerical reasons only three vibrational modes could be considered in
the quantal calculations).

References

1) K. Freed, Topics Appl. Phys. 15 (1976) 23

2) E.J. Heller, J. Chem. Phys. 68 (1978) 2066, 3891

3) H. Köppel, Chem. Phys. 77 (1983) 359

4) H.-D. Meyer, Chem. Phys. 82 (1983) 199

5) H.-D. Meyer and H. Köppel, J. Chem. Phys. 81 (1984) 2605

6) H. Köppel and H.-D. Meyer, Chem. Phys. Lett. 107 (1984) 149

7) H.-D. Meyer and W.H. Miller, J. Chem. Phys. 70 (1979) 3214,
    71 (1979) 2156, 72 (1980) 2272

MOTION OF WAVEPACKETS-SENSITIVITY TO THE INITIAL CONDITIONS

N. Moiseyev

Department of Chemistry, Technion-Israel Institute of Technology,
32000 Haifa, Israel

1. Introduction

If $\hat{H}(r)$ is the molecular time independent Hamiltonian then the energy levels, E_j, and eigenfunctions ψ_j can be obtained by solving the time-independent Schrödinger equation. Let's assume that resulting from external perturbation (e.g. short laser pulse or fast molecular collision) the molecule is taken out of equilibrium and at time which is referred to as t=0 the molecular system is described by $\phi_0 = \psi_j$. The time dependent behavior of our system is obtained by $\phi(r,t) = e^{iHt/\hbar} \phi_0 = \Sigma_j \langle \psi_j | \phi_0 \rangle \exp[-iE_j t/\hbar]\psi_j$. If, for example (r) is the distance operator between two atoms in the molecule then: $\bar{O}(t) = \langle \phi(r,t)|\hat{O}|\phi(r,t)\rangle$ is the bond length as function of time during the unimolecular reaction process. If $\hat{O}$ is a specific molecular mode hamiltonian, e.g. $\hat{O} = \frac{1}{2}(P_\xi^2 + \omega^2 \xi^2)$, $\xi = f(r_1, r_2, \ldots)$ then the energy in this specific mode as function of time will be described by $\bar{O}$. However, in order to get the desired information on the dynamicsl process we should answer the two following questions:

1. What is ϕ_0 ?
2. How to calculate $\phi(t)$?

The difficulty in calculating $\phi(t)$ arises from the fact that we would like to avoid the necessity of calculating ψ_j. See for example the computational methods developed recently by Heller[1]. Since we cannot give a general answer to the first question we shall try to see <u>how sensitive the time dependent behavior of the molecular system is to the initial contions, and if there is any situation in which is not "worth" trying to find ϕ_0</u> (i.e. high sensitivity to the initial conditions).

2. Sensitivity to the initial conditions [2]

If the evolution in time of the initial wavepacket is

$$\phi(r,t) = \sqrt{\rho(r,t)} \ e^{iS(r,t)/\hbar} \tag{1}$$

then by substituting ϕ into the time-dependent Schrödinger equation we get

$$\frac{\partial \rho}{\partial t} + \vec{\nabla}(\rho \ \frac{\Delta \vec{S}}{m}) = 0 \tag{2}$$

$$\frac{\partial S}{\partial t} + \frac{1}{2m} \vec{\nabla} \vec{S} . \vec{\nabla} \vec{S} + V(\vec{r}) = \frac{\hbar^2}{2m} \frac{\nabla^2 \sqrt{\rho}}{\sqrt{\rho}} \ . \tag{3}$$

If $\hbar = 0$ we get from Eq. (2,3) the Hamilton Jacobi classical equation of motion, where ρ is the probability density of the fluid which flows with velocity $\frac{\nabla \vec{S}}{m}$. As we know there are essentially <u>two</u> different types of orbits for classical Hamiltonian systems.

(1) "regular" - <u>not sensitive</u> to the initial conditions, i.e. neighboring
orbits separate at a rate which is (roughly) linear in time;
$\frac{\Delta x(t)}{\Delta x(0)} \sim t.$

(2) "irregular" or "chaotic" - <u>very sensitive</u> to the intial conditions,
$\frac{\Delta x(t)}{\Delta x(0)} \to e^{+\lambda t}$ as $t \to \infty$ and $\Delta x(0) \to o.$

If we neglect $\hbar$ term in Eq. (3) we expect $\phi(r,t)$ to spread a the <u>same</u> rate as a classical ensemble of particles for which the following variances,

$$(\Delta y)^2 = \overline{y^2} - (\bar{y})^2 \quad , \quad (\Delta x)^2 = \overline{x^2} - (\bar{x})^2$$

$$\tag{4}$$

$$(\Delta P_y)^2 = \overline{P_y^2} - (P_y)^2 \quad , \quad (\Delta P_x) = \overline{P_x^2} - (\bar{P}_x)^2$$

increase <u>slowly</u> (linearly) in time in the <u>regular</u> region and, increase <u>rapidly</u> (exponentially) in time in the <u>chaotic</u> region. In quantum mechanics $\hbar \neq 0$. However, we <u>assume</u> that $\phi(r,t)$ will <u>still</u> spread at a <u>similar rate</u> as classical ensemble of particles. Therefore,

$$(\Delta y)^2 = \langle \phi(r,t)|y^2|\phi(r,t)\rangle - \langle \phi|y|\phi\rangle^2$$

$$\tag{5}$$

$$(\Delta P_y)^2 = \langle \hat{P}^2 \rangle - \langle P_y \rangle^2 \text{ and etc.}$$

will increase very rapidly (slowly) if $\phi(t=0)$ is located in the chaotic or (regular) region. We checked the above conjecture by carrying out a numerical experiment. For the Henon-Heiles system, $\hat{H} = \frac{1}{2}(P_x^2 + P_y^2 + x^2 + y^2) + x^2 y - \frac{1}{3}y^3$ (which qualitatively describes coupling between symmetric and antisymmetric vibrational modes in three atomic linear molecules) two different initial conditions were taken. In the first one, $\phi_1(t=0) = (\frac{1}{\pi\hbar})^{1/2} e^{((x-x_0)^2 + (y-y_0)^2)/2\hbar}$. $e^{(i\dot{x}_0 x + i\dot{y}_0 y)/\hbar}$ where, x_0, y_0, $\dot{x}_0$, $\dot{y}_0$ are the classical initial condition of a stable periodic orbit of the Henon-Heiles at E = 0.125 and with 6.083 periodicity.

In the second chosen initial condition the center of ϕ_0 is on an unstable periodic orbit at the same energy at E = 0.125 and with a similar periodicity (τ = 6.908).
In the first case we get,

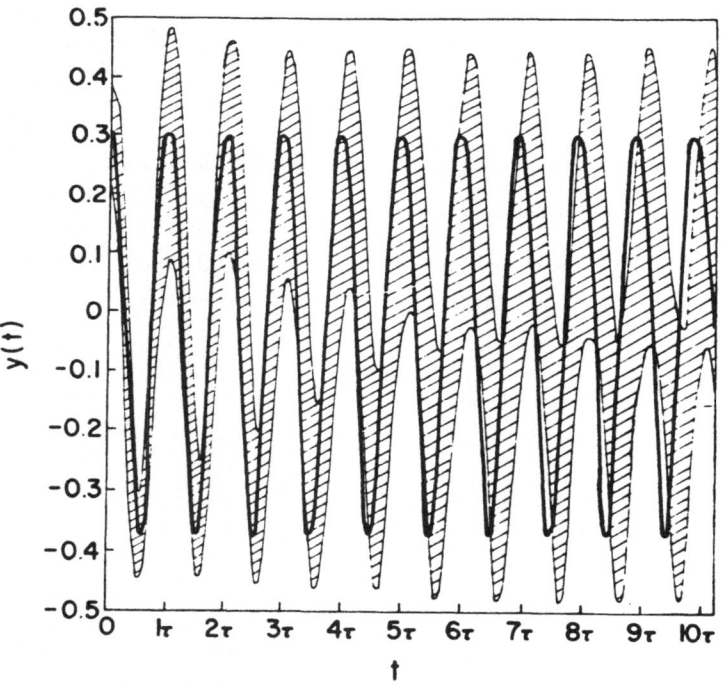

we see that y(t) spread very slowly in time, the quantum uncertainty is small (the dashed area in the above figure) and even after many periods still the molecule "remembers" the initial condition. Note that the quantum periodicity is not exactly the same as the classical one (the dark line) as could be expected, since the quantum wavepacket can be described (in the classical limit) as ensemble of particles. Whereas in the second case one can see from the results presented below, (next page) that the quantum uncertainty is large (the dashed area) and very rapidly (after few oscillation) the molecule <u>does not "remember" the initial conditions and it is not worth trying to</u> <u>calculate ϕ_0 and statistical approaches (like RRKM) should be used</u>. Note that after4-6 periods of oscillations $\phi(t)$ fill up the whole phase space and a standing wavepacket is obtained.

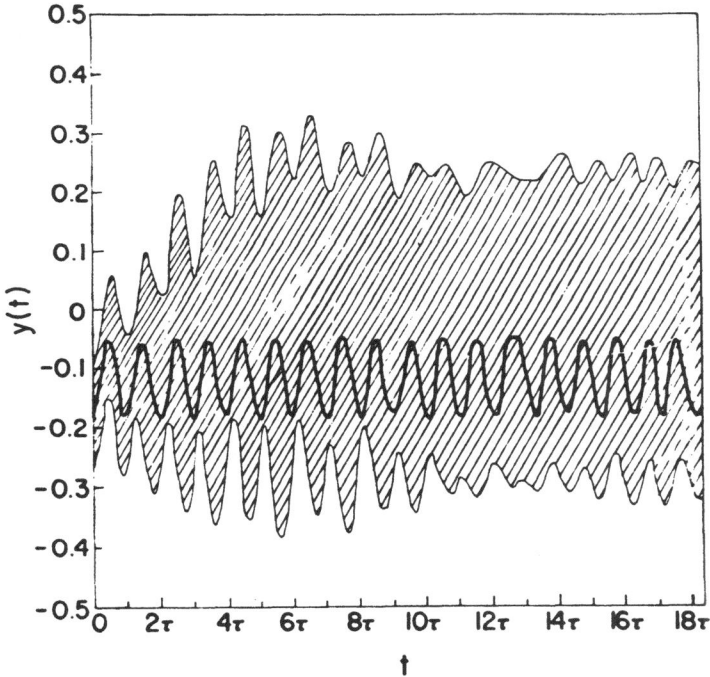

The explanation to the different behavior of the two wavepackets based on classical mechanics. What is the quantum mechanical explanation ? The two wavepackets have the same energy, the same width, and the only possible explanation is that they are constructed from different eigenfunction of $\hat{H}$. The expectation value, $\bar{O}(t) = [<\psi_\ell|\phi_0><\phi_0|\psi_{\ell'}><\psi_{\ell'}|\hat{O}|\psi_\ell> \exp i(E_\ell - E_{\ell'})t/\hbar]$ will show a periodic behavior if $E_\ell - E_{\ell'} = \Delta \ell \hbar \omega$. Namely, $<\psi_\ell|\phi_0> \neq 0$, for $\ell = \ell_1, \ldots$ if $\{E_{\ell 1}, E_{\ell 2}, \ldots\}$ are nearly equidistant. Indeed this is the result obtained in our calculations which are presented in the next figure.

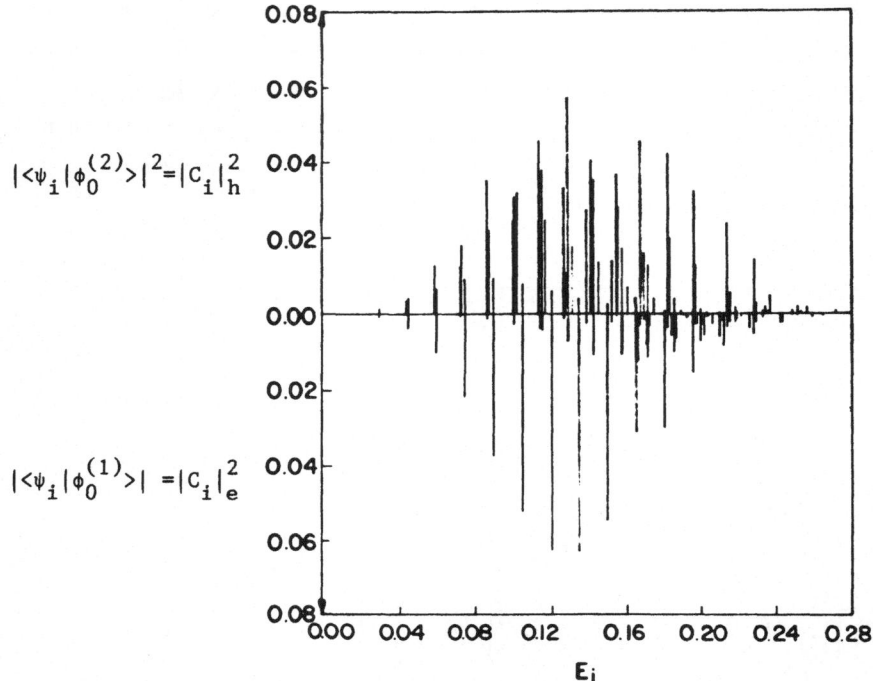

Note, that as proposed by Berry [3], different eigenfunction ("regular" and "irregular") construct ϕ_1, and ϕ_2. In such a case the overlap S,

$$S = |\langle\phi_1|\phi_2\rangle|^2 = \sum_n c_{1n}^* \, c_{2n}$$

$$(6)$$

$$C_{1n} = \langle\psi_n|\phi_0^{(1)}\rangle \quad \text{and} \quad C_{2n} = \langle\psi_n|\phi_0^{(2)}\rangle$$

should vanish as $\hbar \to 0$. The results presented in the next page confirm this analysis.

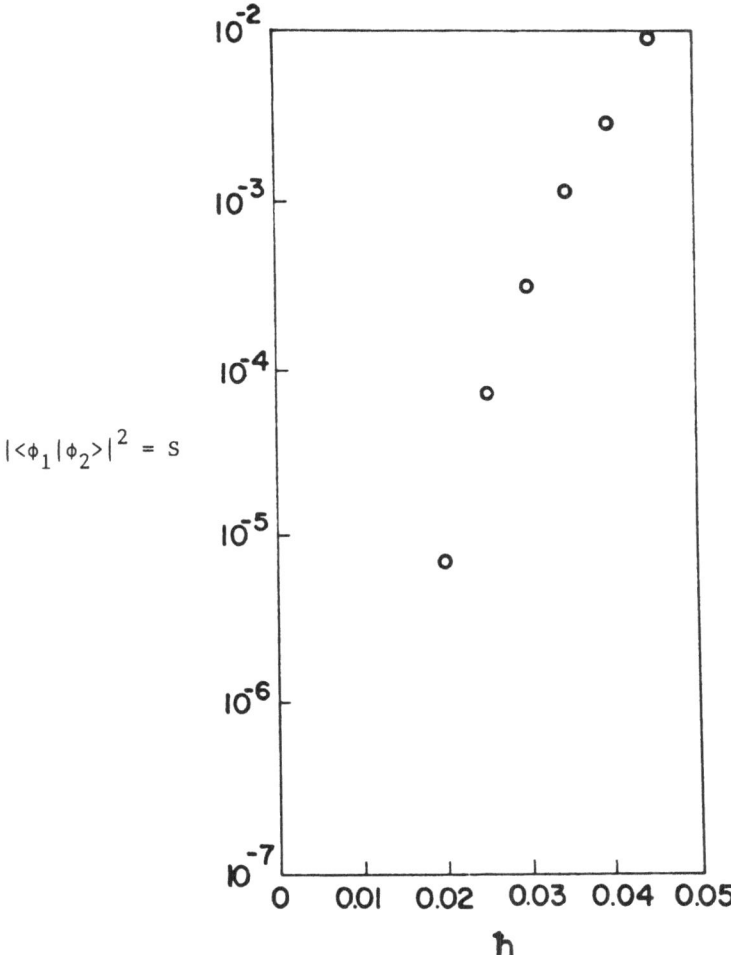

$$|<\phi_1|\phi_2>|^2 = S$$

3. Sensitivity to the initial condition integrable systems (4)

The following two Hamiltonians are non-linear exactly soluble systems (in quantum mechanics) and integrable (in classical mechanics)

$$H_1 = a^+a + b^+b + \frac{1}{2}\sqrt{\frac{\hbar}{2}} [(a^+)^2 b + B^+ a^2]$$

$$H_2 = a^+a + 2b^+b + \frac{1}{2}\sqrt{\frac{\hbar}{2}} [(a^+)^2 b + b^+ a^2]$$

(7)

where a and b are two Boson operators. The <u>exact</u> eigenfunctions of these Hamiltonians are constructive of <u>finit</u> number of harmonic oscillators basis functions, $|n_a\rangle|n_b\rangle$.

Such that

$$H = \begin{pmatrix} \boxed{2\times2} & & 0 \\ & \boxed{3\times3} & \\ 0 & & \boxed{4\times4} \end{pmatrix} \qquad (8)$$

where for example the 2×2 block is constructed of 2 basis functions $|0\rangle|2\rangle$; $|2\rangle|0\rangle$ the 6×6 block from 6 basis functions, $|0\rangle|5\rangle$; $|2\rangle|4\rangle$; $|4\rangle|3\rangle$; $|6\rangle|2\rangle$; $|8\rangle|1\rangle$; $|10\rangle|0\rangle$.

If $\Phi_0 = |90\rangle|0\rangle$ then in the first case (H_1-Hamiltonian) the system has a long memory and even after 40 oscillations the mode "a" is highly excited whereas the mode "b" is not. However, this is not the situation in the second case (H_2-Hamiltonian) where after 6 oscillators we get an equal (<u>statistical</u>) distribution of the energy between the two modes. (See the next Figure).

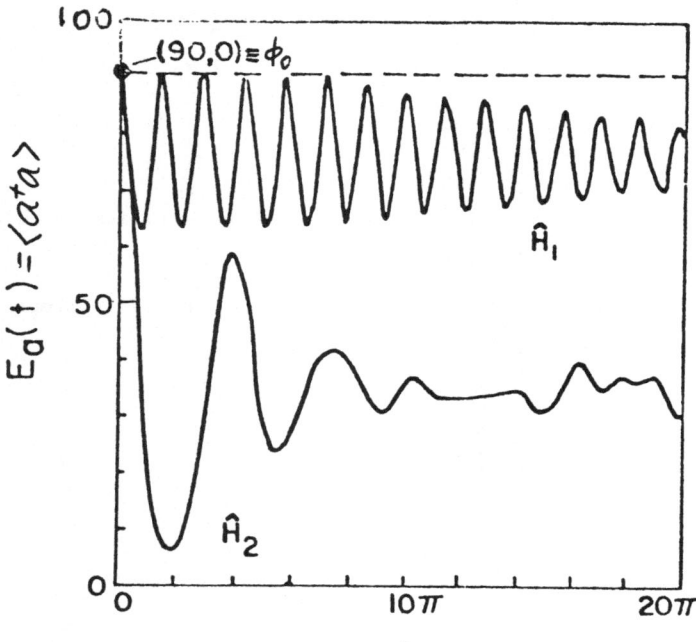

The explanation was given in a work with Katriel (see Figs. 3 in Ref. 5).
It seems that we have an example in which the statistical behavior of the
system cannot be predicted from the classical results. However, the above
Hamiltonian shows a regular behavior after many oscillators (τ is extremely
large) and for any practical time these Hamiltonians behave as chaotic ones.

4. "Regular" and "irregular" energy eigenfunctions (6)

As shown in Section 2 the eigenfunctions of H can be divided into two
classes "regular" and "irregular" eigenfunctions: A wavepacket which is not so
very sensitive to the initial conditions consists of eigenfunctions which belong
to the "regular" class, whereas a wavepacket which is extremely sensitive to
the initial conditions consists of eigenfunctions from the second
class. How one can distinguish between the two classes of eigenfunctions ?

Percival [7] pointed out that the sensitivity to the initial conditions
can be indicated from the sensitivity of the energy spectrum to a perturbation;
$H = H_0 + \lambda V$. Marcus et al suggested <u>many</u> avoided crossings as an indicator to
quantum chaos [8]. To illustrate it we shall make a picture of an isolated
avoided crossing

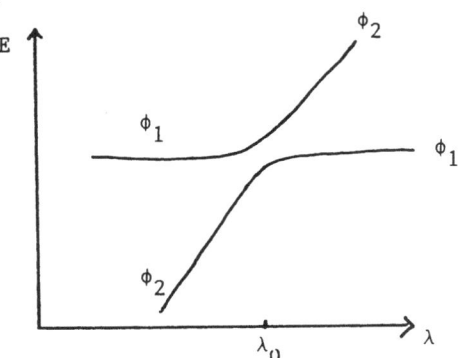

Within $\lambda_0 \pm \Delta\lambda$ $\phi_1 \sim \phi_2$ and are considered as chaotic functions. The question
is, for what value of λ the regular/chaotic transitions occurs ?
A possible answer to this question is as follows:
If, $[\hat{H}_0, \hat{V}] \neq 0$ one can find at least one value of λ, $\lambda_b = |\lambda_b| e^{i\varphi}$ such that,
<u>has as incomplete spectrum</u>. [9,10]
Namely

$$
\begin{aligned}
E_1 &\to E_2 \\
&\qquad \text{as} \qquad \lambda \to \lambda_b \\
\psi_1 &\to \psi_2
\end{aligned}
\tag{9}
$$

and

$$\lim_{\psi_1 \to \psi_2} \int \psi_1 \psi_2 \, d\tau = 0 \tag{10}$$

Note, that since $\hat{H}(\lambda_b)$ is a non-hermitian operator (complex symmetric matrix in the finite matrix approach), the complex product $\langle f^* | f \rangle$ rather than the scalar one $\langle f | f \rangle$ should be used.

We propose here that ψ_b should be considered as an "irregular " or "chaotic" eigenfunction:

(1) $\int \psi_b^2 d\tau = 0$, and therefore ψ_b is a zero correlated function. It is almost impossible to calculate ψ_b. (Note that Shapiro et al indicator for chaos is

$$\int \phi(r)\phi(r+\Delta r)dr \approx 0 \quad , \quad \Delta r \geq \Delta r_0 \,)^{11)} \quad . \tag{11}$$

(2) Even if $\langle 0 \rangle = \langle \psi_b | 0 | \psi_b \rangle / \langle \psi_b | \psi_b \rangle = $ Cor then it is most likely that

$$(\Delta 0)^2 = \langle 0^2 \rangle - \langle 0 \rangle^2 \to \infty \quad \text{as} \quad \lambda \to \lambda_b \quad . \tag{12}$$

We get here that <u>incomplete spectrum implies infinitely large quantum uncertainty</u>.

(3) E does not have a Taylor expansion in $(\lambda-\lambda_b)$ around λ_b, but do have an expansion in $(\lambda-\lambda_b)^{1/p}$ (6) ,

$$E(\lambda) = \sum_k \alpha_k (\lambda-\lambda_b)^{k/p} \tag{13}$$

where p is the number of eigenfunctions that coallesed. Usually (Byeress-Brown, Stillinger) [11,12)

$$E(\lambda) \sim [(\lambda-\lambda_b^*)(\lambda-\lambda_b)]^{1/2} \tag{14}$$

and therefore

$$\left. \frac{\partial E}{\partial \lambda} \right|_{\lambda b} = \infty \tag{15}$$

This is a clear indicator for chaos as proposed by Percival [7].

Numerical calculations [6] for two dimensional model Hamiltonians have shown that in the energy-interval for which most of the corresponding phase space is regular $|\lambda_b| \gg |\lambda$ (the physical value)$|$. In the energy-interval however for which almost all the corresponding phase space is irregular then $|\lambda_b| \simeq |\lambda|$ and the eigenfunction of the Hermitian Hamiltonian are "almost" self orthogonal and Eqs. 12,15 are almost satisfied. Following the above definition of irregular eigenfunction it is obvious that a clear distinction between regular and irregular wavefunction can be made only for complex non hermitian (perhaps non-physical) Hamiltonian irregularity. Thus in the strict mathematical sense all the spectrum of a physical Hamiltonian is regular. However, if the spectrum of the analytically continued hamiltonian includes an irregular eigenvalue close to real axis as a significant part of its irregularity is projected onto corresponding eigenvalue of the physical hamiltonian.

References

1) E.J. Heller, J. Chem. Phys. 62 (1975) 1544; Chem. Phys. Lett. 34 (1975) 321; J. Chem. Phys. 64 (1976) 63; ibid. 65 (1976) 4979; ibid. J. Chem. Phys. 68, (1978) 2066; ibid, 75 (1981) 2923.

2) N. Moiseyev and A. Peres, J. Chem. Phys. 79 (1983) 5945

3) M.V. Berry, Philos. Trans. R. Soc. London, Ser.A287 (1977) 237

4) N. Moiseyev, J. Phys. Chem. 87 (1983) 3420

5) J. Katriel and N. Moiseyev, J. Chem. Phys. 78 (1983) 876

6) N. Moiseyev, to be published.

7) I.C. Percival, J. Phys. B6 (1973) 1229

8) R.A. Marcus, in Quantum Chaos, edited by G. Cassati and J. Ford (Academic, New York 1984), R. Ramaswamy and R.A. Marcus, J. Chem. Phys. 74 (1981) 1379; ibid, (1981) 1385; D.W. Noid, M.L. Koszykowski, M. Tabor and R.A. Marcus, J. Chem. Phys. 72 (1980) 6169

9) T.S. Motzkin and O. Taussky, Trans. Am. Math. Soc. 80 (1955) 387

10) N. Moiseyev and S. Friedland, Phys. Rev. A22 (1980) 618

11) M. Shapiro and O. Goelman, Phys. Rev. Lett. 53 (1984) 1714

12) W. Byers Brown, unpublished, see N. Moiseyev and P.R. Certain, Mol. Phys. 37 (1979) 1621.

13) F. Stillinger, J. Chem. Phys. 45 (1966) 3623

QUANTUM RESONANCES
IN PHYSICAL TUNNELING

M.M. Nieto

Theoretical Division, Los Alamos National Laboratory,
University of California, Los Alamos, NM 87545, U.S.A.
The Niels Bohr Institut, Blegdamsvej 17, 2100 København ø, Denmark[*]

D.R. Truax [**]

Theoretical Division and Center for Nonlinear Studies
Los Alamos National Laboratory
University of California, Los Alamos, NM 87545, U.S.A.

It has recently been emphasized that the probability of quantum tunneling
is a critical function of the shape of the potential. Applying this observation to physical systems, we point out that in principal information on
potential surfaces can be obtained by studying tunneling rates. This is especially true in cases where only spectral data in known, since many potentials
yield the same spectrum.

Historically, our intuition on tunneling comes mainly from two places.
The first is from alpha decay [1]. There we know that the nucleus eventually
decays; it is a resonance coupled to the continuum. The second source of our
intuition is from symmetric double-well potentials. There it is well known [2]
that the ground and first excited states are almost degenerate. The ground
state has a symmetric wave function (no nodes) and the first excited states
has an antisymmetric wave function (one node). The time it takes for a wave
packet originally located on one side to get to the other side is given by
$\tau = \pi\hbar/(E_1 - E_0)$. The wave packet tunnels back and forth in multiples of this
oscillation time.

From the above intuition it might be assumed that given any double-well
potential the "particle" will always tunnel to the lower well, given enough
time. However, in general this is not true. Whether a state can signi-
ficantly tunnel to the true vacuum is a very sensitive function of the shape
of the potential.

---

[*] Address during the academic year 1985-86.

[**] Returning Aug. 1985, from leave of absence to the Department of Chemistry,
University of Calgary, Calgary, Alberta, Canada T2N 1N4

This was recently emphasized in work [3] which was motivated by a very different type of study, the early universe. Starting from the dimensionless Schrödinger equation

$$[- \frac{d^2}{dx^2} + V(x)]\Psi = i \frac{d}{dt} \Psi, \tag{1}$$

this group studied tunneling from the higher left-hand well in the potentials

$$V = \begin{cases} (x+2)^2 + U & x \leq 0 \\ (x-[4+U]^{1/2})^2, & x \geq 0 \end{cases} \tag{2}$$

$$V = \begin{bmatrix} \frac{5}{(18)^2} (x+3)^2(x-2)(x-6), & x \leq \frac{9}{2} \\ -\frac{(25)^2}{(12)(16)} & \frac{9}{2} \leq x \leq d + \frac{9}{2} \\ \frac{5}{(18)^2} (x-d+3)^2(x-d-2)(x-d-6), & d + \frac{9}{2} \leq x, \end{bmatrix} \tag{3}$$

and

$$V = \begin{cases} 6-x & x \leq -4 \\ 0, & -4 \leq x \leq -1 \\ 2, & -1 \leq x \leq 1 \\ -U, & 1 \leq x \leq 4 \\ 6+x, & 4 \leq x \end{cases} \tag{4}$$

These potentials are shown in figs. 1, 2 and 3 respectively.

134

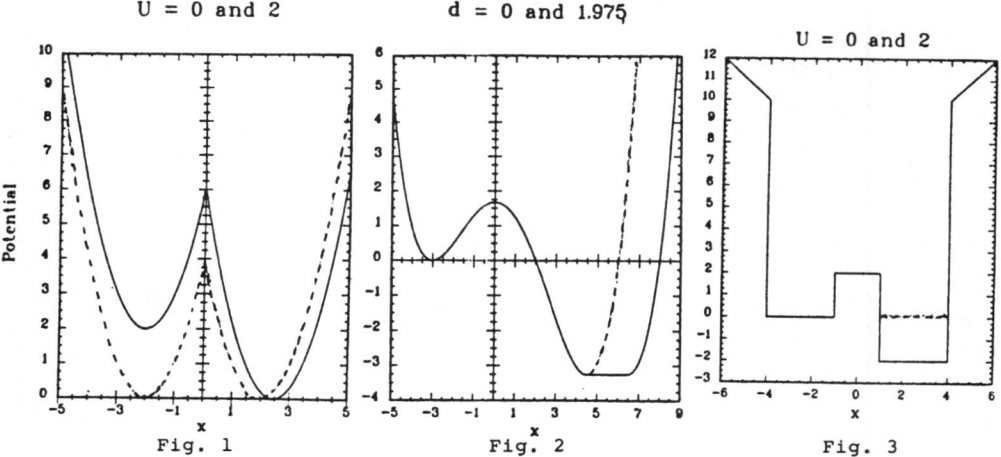

U = 0 and 2   d = 0 and 1.975   U = 0 and 2

Fig. 1   Fig. 2   Fig. 3

The true minimum of the potential of Eq. (2) lowers and goes to the right as U is increased. The true minimum of the potential of eq. (3) expands to the right as d increases, and the true minimum of the potential of Eq. (4) lowers as U increases. In all three potentials a gaussian wave-function, fit to the false minimum, was placed in the left-hand well. Then, in all three cases, the wave-function was numerically integrated in time. In the first case an analytic check was done which agreed with the numerical integration.

Using these results, the maximum probability that the false-vacuum state ever penetrates to the right-hand side, for any time, for arbitrary U (or d) was plotted. This probabilty is defined as

$$\rho_{RHS}^{max}(U) = \max_{\forall t} \int_0^\infty dx\ \Psi_0^*(x,t)\Psi_0(x,t).$$ (5)

In figs. 4-6 are shown the results for the three potentials.

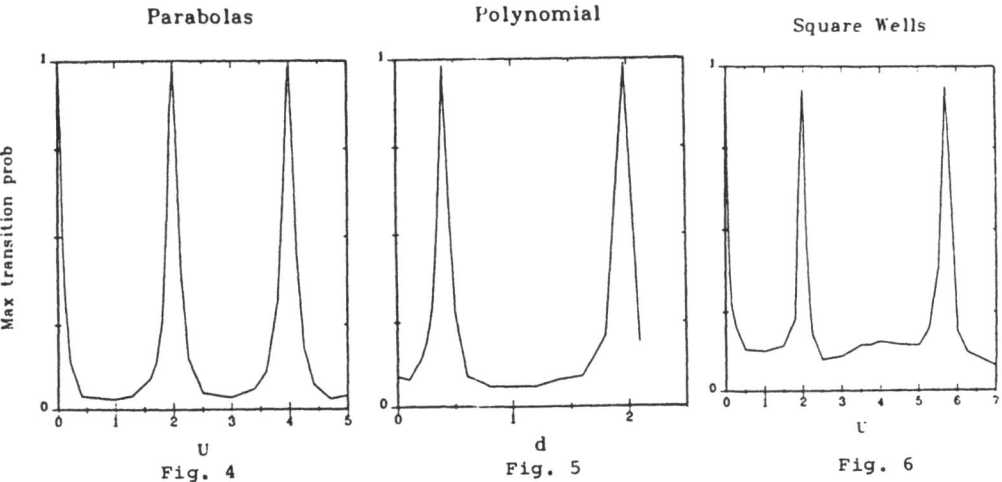

Parabolas   Polynomial   Square Wells

Fig. 4   Fig. 5   Fig. 6

A line plot of the maximum probability which can tunnel from the false vacuum
for the potentials in Eq. (2), in Eq. (3), and Eq. (4). The time evolutions
of the wave functions ware calculated for times as large as 300. These plots
have an accuracy of a few percent. The tunneling times at the maxima are the
following: F.4 τ = 38,28 and 25 for the maxima at U = 0,2 and 4; F.5 $\tau \cong$ 58
and 69 for the maxima at d = 0.4 and 1.975; F.6 τ = 44, 32 and 30 for the
maxima at U = 0, 1.98 and 5.7. Observe that even at the resonance peaks the
maximum probability that tunnels to the true vacuum is not 100%. This is
because there always is some fraction of the wavepacket that remains to the
left of x = 0.

To understand what is going on, look at the first three eigenfunctions
of Eq. (2) for U = 0, 1 and 2 (Figs. 7, 8 9).

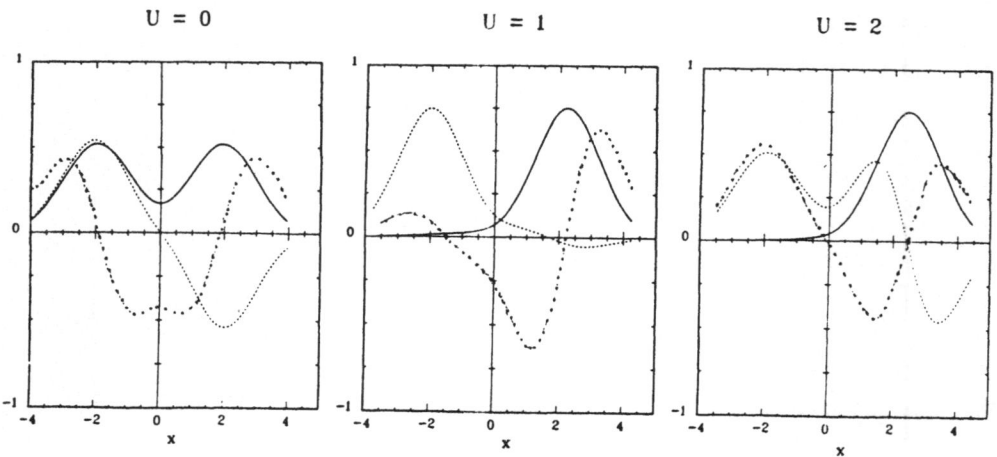

Figs. 7-9: The first three eigenvectors ϕ_0, ϕ_1 and ϕ_2 of Eq. (2) for (7) U = 0, (8) U = 1, and (9) U = 2. The ground state ϕ_0 is represented by the solid line, ϕ_1 by the close-dashed line, and ϕ_2 by the spaced-dashed line.

At U=0, the false vacuum eigenfunction almost equally overlaps with ϕ_0 and ϕ_1 on the left-hand side. When the tunneling time τ is reached, these two eigenfunctions are out of phase by π, and so the overlap is on the right-hand side. The overlap of ψ_0 with ϕ_2 is almost zero because of the node of ϕ_2 near x = -2. For U = 1, ψ_0 has an overlap almost exclusively with ϕ_1 which now, however, is almost entirely on the left. Therefore, since ψ_0 has no significant overlap with an eigenfunction with large probability on the right, it can <u>never</u> tunnel to the right. When U = 2, we have a situation similar to U = 0, except that the two almost degenerate eigenstates are now ϕ_1 and ϕ_2. They both have nodes near x = 2.4. This means that when the wave packet tunnels to the right, it <u>will not</u> be a coherent tunneling, but rather will have two bumps.

Clearly this is a resonance condition. It is similar to matching cavities with electromagnetic waves. When the two wells are tuned correctly, the waves can pass back and forth. The tuning is caused by integrating the curvature of the wavefunction properly, this curvature being [E-V(x)]. For the present harmonic oscillator-like potential, this means a jump of U = 2.

Therefore, as the graphs in Figs. 4-6 show, if one is interested in
tunneling, it is important to know how good a fit a particular parametrization
of a double-well potential actually is to the "true" potential. As was shown [3]
small deviations can have drastic effects on the tunneling rates.

Furthermore, since many potential surfaces are determined by the spectra
alone, and not by the scattering data, the picture is more complicated still.
For given only a spectum, there are an <u>uncountable</u> number of potentials which
will give the same spectrum. This fact has been beautifully expounded by
Abraham and Moses [4], and elucidated for different purposes elsewhere [5-7].

Putting the "one-halfs" back in ere to conform to the Fig. from Ref. 5,
we give an example of AB^4. If one adds to the harmonic oscillator potential
v_0 the potential v_1:

$$v = v_0 + v_1, \tag{6}$$

$$v_0 = \frac{1}{2} z^2 \quad , \quad v_1 = 4\phi(\phi - z), \tag{7}$$

$$\phi(z) = \frac{e^{-z^2}}{z^{1/2} \operatorname{erfc}(z)}, \tag{8}$$

$$\operatorname{erfc}(z) = \frac{2}{\pi^{1/2}} \int_z^\infty e^{-t^2} dt, \tag{9}$$

one has an <u>analytically solvable potential</u> with the same spectrum as the
original, but with the ground state removed [4-6]. This is shown [5] in Fig.10

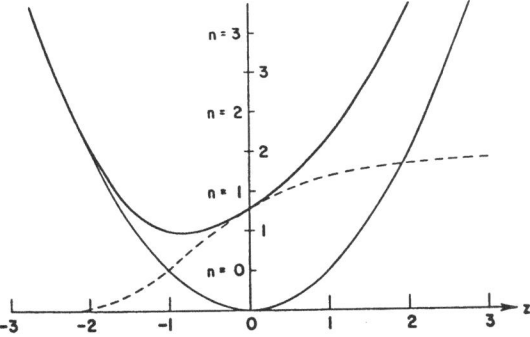

Fig. 10: The harmonic oscillator potential $v_0(z)$ is a light curve, the
contribution $v_1(z)$ is a light dashed curve, and the complete AM
potential $v(z)$ is a heavy curve. The number eigenstates are also
indicated.

Now by simply adding -1 to the potential, one has the original spectrum. This is the simplest example, but the program works in general, including for potentials with a continuum.

With these points in mind, we became interested in applying these insights to physical systems. We considered the work of Wieser and collaborators [8]. From spectra, they obtained phenomenological potentials for the methyl-substituted oxetans and thietans of the form

$$v = N_2 x^2 + N_3 x^3 + N_4 x^4. \tag{10}$$

In table I we give, in the notation of Eq. (1), their values of the N_i for three specific asymmetric compounds.

	2-methyl oxetan	3-methyl oxetan	2-methyl thietan
N_2	-0.6736	-0.5964	-0.862
N_3	-0.0434	-0.05866	-0.0275
N_4	0.07028	0.06856	0.03422
ρ_{MAX}^{RHS}	0.36	0.51	0.02
τ	3.06	3.10	19.18

Table 1

Using the above values for the N_i, we found that the oxetans allowed a significant fraction of the wave-packet to tunnel in a short time whereas the thietan allowed almost no tunneling. This is also shown in Table 1.

We emphasize that the present calculations apply to no dissipation. Thus, we are talking about a situation where very low temperatures and pressures would be involved. The idea is to try to use the calculated tunneling to test the accuracy of phenomenological potentials in the low energy regeme. Hopefully the calculations would be good enough to compare to future experiments.

Cribb and coworkers [9-11] have previously undertaken related tunneling calculations. They used a density matrix formalism in their work. They started [9] by considering four sets of parameters for a specific potential and showed how the tunneling properties varied with the different sets.

They then [10,11] applied their formalism to chemical systems where pressure and temperature effects are important. These are problems of practical interest in standard chemical reactions.

Our focus is different. We are interested in determining the potential energy surfaces in systems where pressure and temperature effects can be ignored. In that case, tunneling may in principle provide a method to distinguish between slightly different phenomenological potentials in the low energy regeme.

Switching quickly to nuclear physics, we mention the observation of Berger at this meeting. In discussing the two-dimensional potential energy surface his group has obtained [12] for the fission of ^{240}Pb, Berger observed [13] that the small amount of low-energy fission component might be due to resonant tunneling. The tunneling would be from the higher-energy valley to the (approximately parallel) lower-energy valley, both of which are approximately perpendicular to the normal escape direction in the two-dimensional surface. This observation is worthy of further consideration.

Acknowledgements

We first wish to thank H. Wieser for comments on his experiments, his phenomenological potentials, and their relationships to tunneling. Among the many participants at the meeting who were kind enough to provide comments and information, special thanks are due P.H. Cribb and J.F. Berger.

References

1) E. Fermi, J. Orear, A.H. Rosenfeld and R.A. Schluter, Nuclear Physics, Revised Ed. (Univ. of Chicago 1950), Ch.III.

2) E. Merzbacher, Quantum Mechanics, Second Ed. (Wiley, New York 19 0) p. 65

3) M.M. Nieto, V.P. Gutschick, C. Bender, F. Cooper and D. Strottman, Phys. Lett. B (to be published).

4) P.B. Abraham and H.E. Moses, Phys. Rev. A22 (1980) 1333

5) M.M. Nieto and V.P. Gutschick, Phys. Rev. D23 (1981) 922;
   M.M. Nieto, Phys. Rev. D24 (1981) 1030

6) B. Mielnik, J. Math. Phys. 25 (1984) 3387

7) M.M. Nieto, Phys. Lett. 145B (1984) 208

8) J.A. Duckett, T.L. Smithson and H. Wieser, J. Molec. Struct. $\underline{56}$ (1979) 157

9) P.H. Cribb, S. Nordholm and N.S. Hush, Chem. Phys. $\underline{44}$ (1979) 315

10) P.H. Cribb, S. Nordholm and N.S. Hush, Chem. Phys. $\underline{47}$ (1980) 135;
    ibid. $\underline{69}$ (1982) 259

11) P.H. Cribb, Chem. Phys. $\underline{88}$ (1984) 47

12) J.F. Berger, M. Girod and D. Gogny, Nucl. Phys. $\underline{A428}$ (1984) 23c

13) J.F. Berger, these proceedings.

WAVEPACKETS IN NUCLEAR PHYSICS

P. Schuck

Institut des Sciences Nucléaires, Université de Grenoble,
53 Avenue des Martyrs, 38026 Grenoble Cedex, France

0) Introduction

Nuclear physics is full of time-dependent processes and semiclassical wavepacket dynamics has of course also been applied to many aspects of nuclear physics. Some problems have been elaborated in quite some detail, some are just at the beginning. Some problems involve purely theoretical and conceptual aspects, some are more oriented towards numerical evaluation of actual problems.

Let me start in recalling some basic concepts of nuclear physics: the nucleus is a <u>selfbound system of Fermions</u> and Hartree-Fock theory is a very valuable first approximation. This stems from the fact that the nucleus is a Fermi system at zero temperature and in such systems two body collisions are very scarce.

The nuclear Hamiltonian is therefore of the single particle type

$$H_\rho = T + V[\rho] \tag{1}$$

where ρ is the single particle density matrix solution of (in the static case):

$$[H_\rho, \rho] = 0 \quad . \tag{2}$$

The corresponding wave functions are Slater determinants.

1) Time Dependent Hartree Fock (TDHF)

Time dependent processes involve deformations of the nucleus. These can e.g. be small amplitude density vibrations or quadrupole shape deformations:

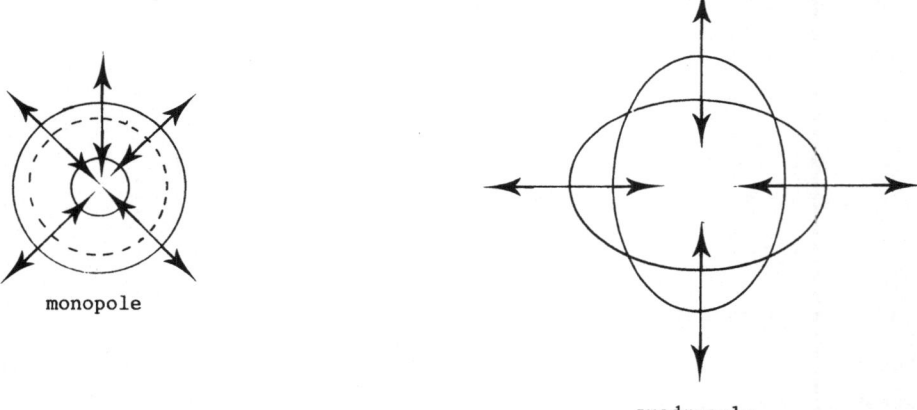

monopole

quadrupole

or large amplitude (collective) motions such as one encounters in nuclear fission or heavy ion collisions. Schematically

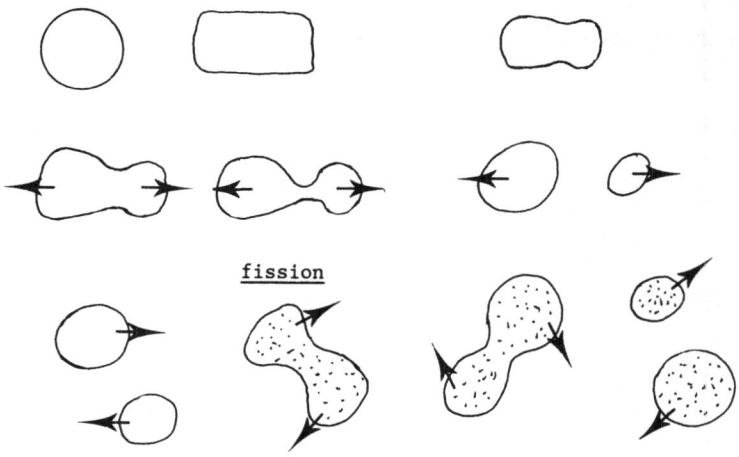

<u>fission</u>

<u>Heavy ion collision with internal heat up.</u>

Evidently for such processes a wave packet description is adequate:

$$|\psi(t)> = |0> + \sum_{\nu \neq 0} c_\nu e^{-\frac{i}{\hbar} \Omega_\nu t} |\nu> \qquad (3)$$

with $\Omega_\nu = E_\nu - E_0$, the excitation energies of the system corresponding to the eigenstates $|0\rangle$, $|\nu\rangle$. Under the assumption that the nucleus stays a Slater determinant for all times if it was initially in a determinal state we then arrive at the time dependent Hartree-Fock equations (TDHF) [1]:

$$i\hbar\dot\rho = [H_\rho, \rho] \tag{4}$$

Many numerical studies of this equation have been performed by now. A typical example is shown in the figure [2].

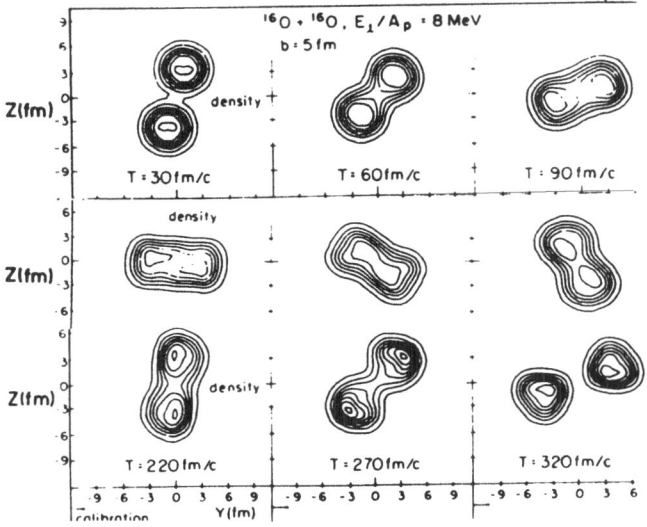

I want to touch a first aspect of this equation which is conceptual. This corresponds to the fact that (4) is in a certain sense a classical approxima-tion. Thouless' theorem tells us that a Slater determinant $\phi(t)$ at time t can be obtained from the static equilibrium solution $|HF\rangle$ by [1,App.E]

$$|\phi(t)\rangle = \exp\left\{\sum_{mi} Z_{mi}(t)\, a_m^+ a_i\right\}|HF\rangle \tag{5}$$

where Z_{mi} are time dependent complex numbers. The total HF energy can be shown to be a constant of motion:

$$\langle\phi(t)|H|\phi(t)\rangle = \text{const.} = E(Z_{mi}) \tag{6}$$

The Z_{mi} form a vector $\vec{t}$ and the vectors $\vec{Z}, \vec{Z}^*$ can be transformed into something like abstract momenta $\vec{P}$ and coordinates $\vec{Q}$. We then have

$$E(\vec{Z},\vec{Z}^*) \rightarrow E(\vec{P},\vec{Q}) \equiv E_{kin} + E_{pot} \quad . \tag{7}$$

The multidimensional Hamilton functions $E(\vec{P},\vec{Q})$ can eventually be separated into a kinetic part E_{kin} and a potential part E_{pot}. In a two dimensional representation we have schematically

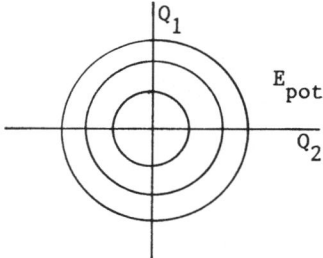

and in the corresponding phase space (in one dimension) the motion corresponds to a line, i.e. the motion in the space of Slater determinants characterized by $\vec{Z}, \vec{Z}^*$ or $\vec{P}, \vec{Q}$ is completely classical.

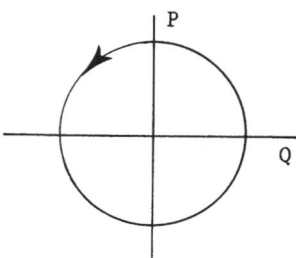

The vectors $\vec{P}, \vec{Q}$ contain collective coordinates (like e.g. deformation) and non collective coordinates (like single particle degrees of freedom). If we represent the fission process by one deformation coordinate Q than in a very simplified form the deformation energy looks like

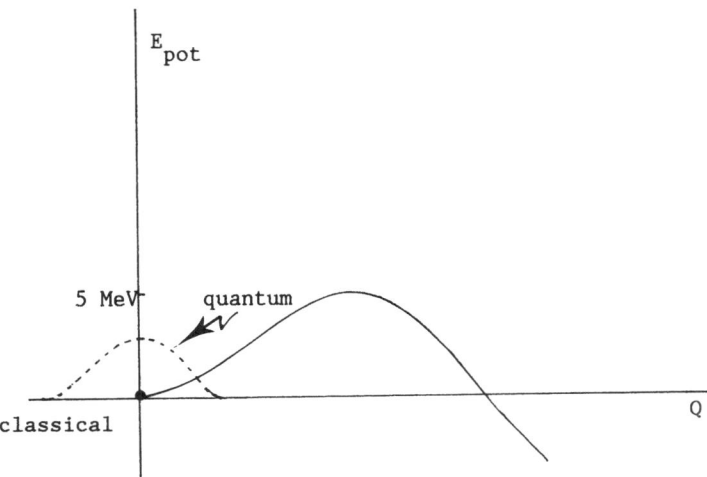

The deformation coordinates represent in nuclear physics what the atomic centers are in molecular physics and a similar Born-Oppenheimer approximation is employed in nuclear physics to calculate such potential energy landscapes [1,3] (adiabatic time dependent H.F.). Since TDHF is a classical theory, if we started our calculation in the minimum of the potential (dot in the figure), we would have

$$ih \, \dot{\rho} = 0$$

forever although quantum mechanically we of course could have tunnelling through the fission barrier.

If we transform our abstract coordinates $\vec{P}, \vec{Q}$ into corresponding abstract phase space variables $\vec{p}, \vec{x}$ then we can represent our system by a phase space distribution and the TDHF solution is given by

$$f_c(\vec{x}, \vec{p}, t) = \delta(\vec{x} - \vec{x}_c(t)) \, \delta(\vec{p} - \vec{p}_c(t)) \tag{8}$$

where the x_c, p_c are the TDHF solutions of the (classical) Hamilton eqs. corresponding to $E(\vec{P}, \vec{Q})$. Formally it is clear how to go beyond TDHF and describe tunnelling: we have to quantise:

$$E(\vec{P}, \vec{Q}) \rightarrow \mathcal{H}(\vec{P}, \vec{Q}) \quad . \tag{9}$$

146

One way to obtain a correct quantised form $\mathcal{H}(\hat{P},\hat{Q})$ is through boson expansion techniques [1,Ch.9]. If the motion is of small amplitude than $\mathcal{H}$ is quadratic in $\hat{P}$ and $\hat{Q}$ and the full quantum solution for any initial Gaussian packet is very well known [4]. In a one dimensional example we have

$$f_c \to f_{QM} = c \exp \{ - \frac{1}{4\Delta}[D_{xx}(p-p_c)^2 - 2 D_{xp}(x-x_c)(p-p_c) + D_{pp}(x-x_c)^2]\} \quad (10)$$

$$\Delta = D_{xx} D_{pp} - D_{px}^2$$

where the D's are the (time dependent) variances and co-variances. In phase space our classical line is now smeared out somewhat due to the finite values of the variances.

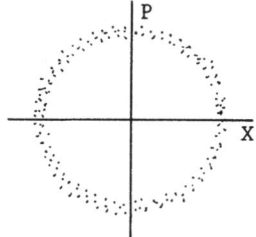

The quadratic approximation is called the random phase approximation (RPA) [1,Ch.8]. As I said these quantised forms go beyond time dependent mean field theory. There have also been attempts to derive them from Feynman path integral techniques [5]. At present studies are under way to quantise in finding closed orbits in the multi dimensional surfaces (Griffin Kit-Keung-Kan; and Ploszajczak, the conference). Another way to describe tunnelling is to solve TDHF with imaginary times ($t \to i\tau$). Negele [6] and Reinhardt [7] have proposed this way.

In many situations the collective coordinates couple directly to the non collective intrinsic coordinates of the nucleons; this gives rise to dissipation of energy of the collective coordinates transformed into internal heat. This is taken account of by Fokker Planck like equations (developed for instance by Hofmann and Siemens [8] and others [9]). Indeed it can be verified that (10) is solution of the following equation:

$$\dot{f} + \frac{\vec{p}}{m} \vec{\nabla}_x f - \frac{\partial E_{pot}}{\partial \vec{x}} \cdot \frac{\partial f}{\partial \vec{p}} = \frac{\gamma}{M} \frac{\partial}{\partial \vec{p}} (\vec{p}-f) + D \frac{\partial^2 f}{\partial p^2} \qquad (11)$$

with $D = \gamma T$ the diffusion coefficient and

$$\frac{\partial E_{pot}}{\partial x} = E'_{pot}(x_c) + E''_{pot}(x_c)(x-x_c) \quad .$$

With the left hand side of (11) included (Fokker Planck) (10) is still solution but the x_c, p_c then obey classical eqs. of motion with friction terms.

A typical example where wuch an equation can be used is represented in the Figure where in a heavy ion reaction nucleons are exchanged between the two ions during the collision process. The charge distribution of the lighter ion is plotted for different kinetic energy losses. The greater the energy loss the wider is the dispersion. This stems from the fact that more energy loss reflects a longer interaction time and thus more particles can be exchanged. Schematically the potential landscape as a function of deformation δ and mass asymmetry A can be depicted as follows

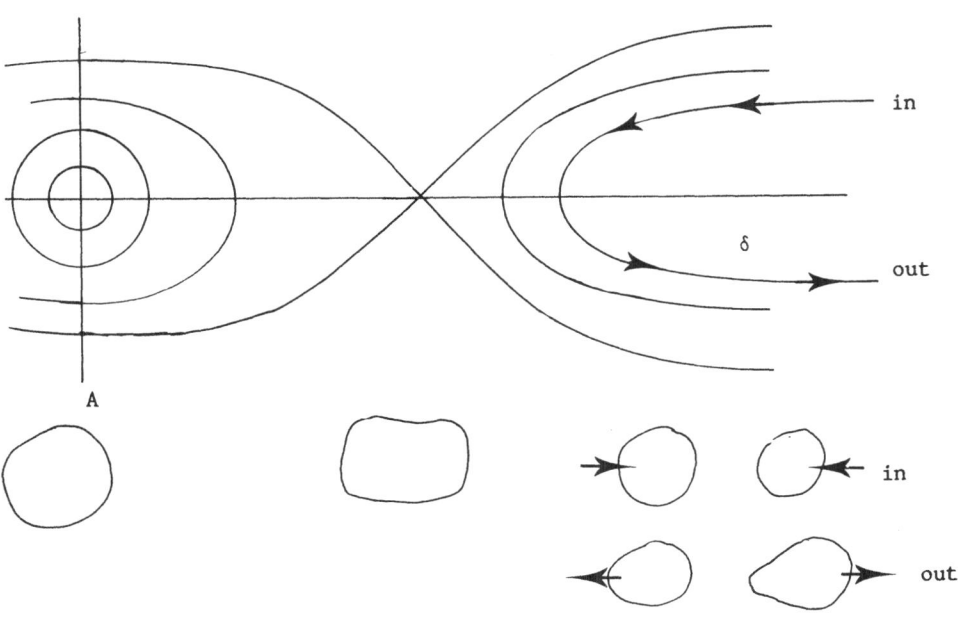

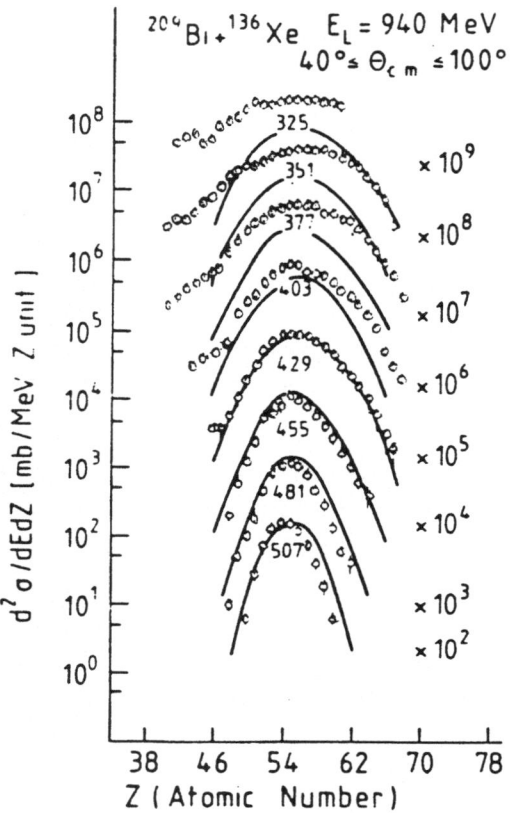

$^{204}Bi + {}^{136}Xe \quad E_L = 940 \text{ MeV}$

$40° \leq \Theta_{c.m.} \leq 100°$

In such a landscape equation (11) has been solved with the ansatz (10) of a
coherent state and the solution is shown as full lines on the Figure.
This discussion on formal aspects of TDHF should be sufficient for an introduction. A much more elaborate example of application will be given by
F. Berger (this conference).

2. Multiphonon states in nuclei.

In very peripheral heavy ion collisions it can happen that the
nucleus is excited into a strong vibrational coherent state. The matter is still
subject of some controversy but there are strong experimental and theoretical
indications that this is actually the case [10,11]. In any case this is a
specially beautiful example of a realisation of a coherent state in nature.
The experimental situation is as shown in the graph:

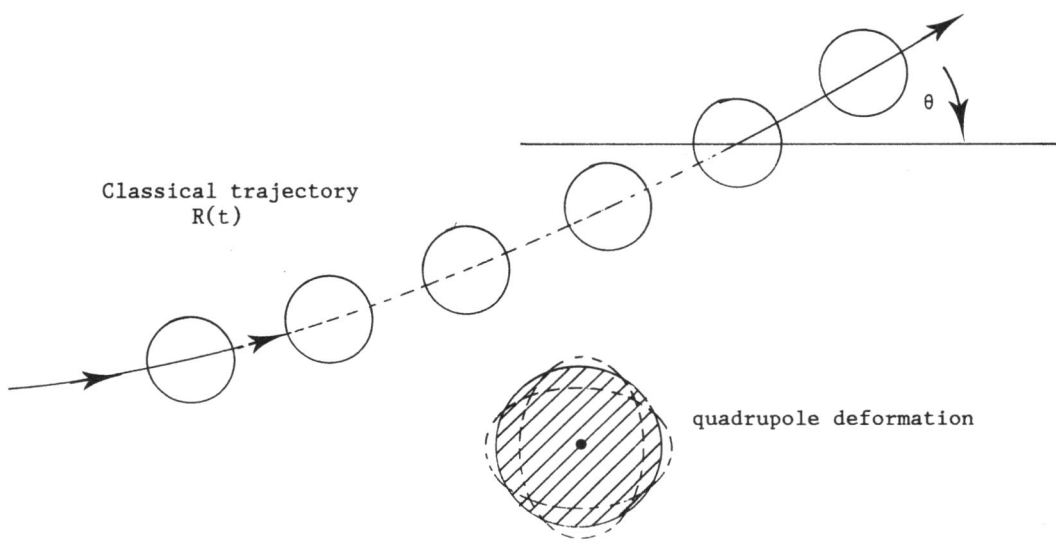

Classical trajectory
R(t)

quadrupole deformation

The closest approach of the two nuclei is such that only the "tails" of their mutual density distributions are overlapping. Furthermore the target nucleus (e.g. ^{208}Pb) is quite stiff against shape deformations so that a harmonic approximation to the Hamiltonian (9) is good enough:

$$\mathcal{H}(\hat{P},\hat{Q}) = \sum_\nu \Omega_\nu \, b_\nu^+ \, b_\nu \tag{12}$$

where the Ω_ν, b_ν^+ are the phonon (vibration) energies and creation operators respectively. If the projectile is represented by a moving Woods Saxon potential

$$V(t) = \frac{V_0}{1 + e^{(R-R_c(t))/a}} \tag{13}$$

where $R_c(t)$ is the classical trajectory of the projectile then to lowest order in the projectile target coupling we have the following time dependent problem describing this particular heavy ion reaction

$$H(t) = \sum_\nu \Omega_\nu \, b_\nu^+ \, b_\nu + \sum_\nu \left[<0|V(t)|\nu> \, b^+ + h.c. \right] \tag{14}$$

$$i\hbar \, \dot{\phi}(t) = H(t)\phi(t) \tag{15}$$

The meaning of $|0>$, $|v>$ in (14) is the groundstate and excited state wave functions of the target nucleus respectively.

The solution of the problem (14,15) is well known [4] to be a quasi classical coherent state of the form

$$|\phi(t=-\infty)> = |0>$$

$$|\phi(t)> = \prod_{v} \exp [A_v(t)b_v^+ - A_v^*(t)b_v]|0>$$

$$\propto \sum_{v,n} C_{v,n}(t)(b_v^+)^n |0> \qquad (16)$$

with

$$A_v(t) = \frac{1}{\hbar} \int_{\infty}^{t} dt' <v|V(t')|0> e^{i\Omega_v t'}$$

From this we can calculate the spectral density (probability) for the excitation of n_v phonons in the state v:

$$P(E) = \sum_{v} |A_v(+\infty)|^2 \delta(E-\Omega_v) \qquad (17)$$

This quantity enters directly into the expression for the cross section $d\sigma/dE$. The excitations $|v>$ have been obtained from microscopic RPA calculations (N.v. Giai) and it turned out that mostly the giant quadrupole resonance $|v> = |2^+>$ contributes to the sum in (17). The state (16) represents then an excitation of a multi 2^+ phonon state. Graphically this can be illustrated as a successive excitation of 2^+ phonon as the projectile ($V(t)$) passes by exciting in the target nucleus many (correlated) particle-hole states. On the experimental side the investigations have been initiated by N. Frascaria et al. [10] and the above presented theory has been developed by Ph. Chormaz and D. Vautherin [11]; their theory is similar to an earlier one by Broglia and Winther [12].

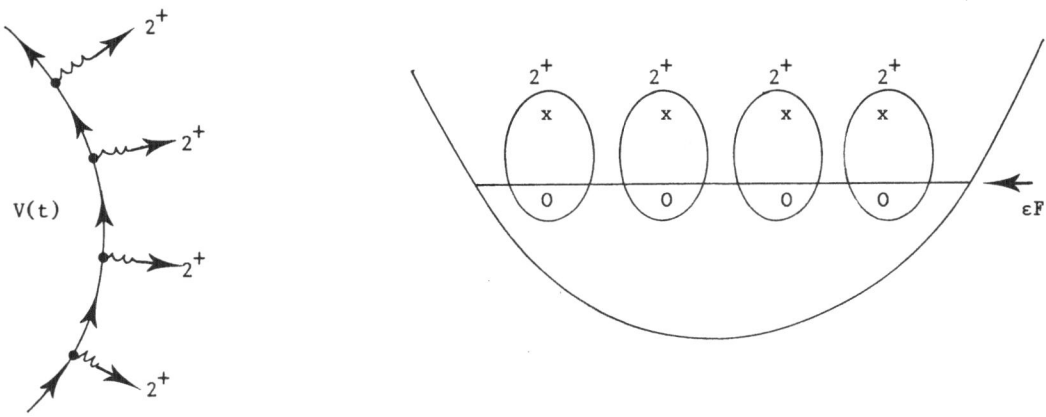

On the figure we show the comparison between experiment (rhs) and theory (lhs)

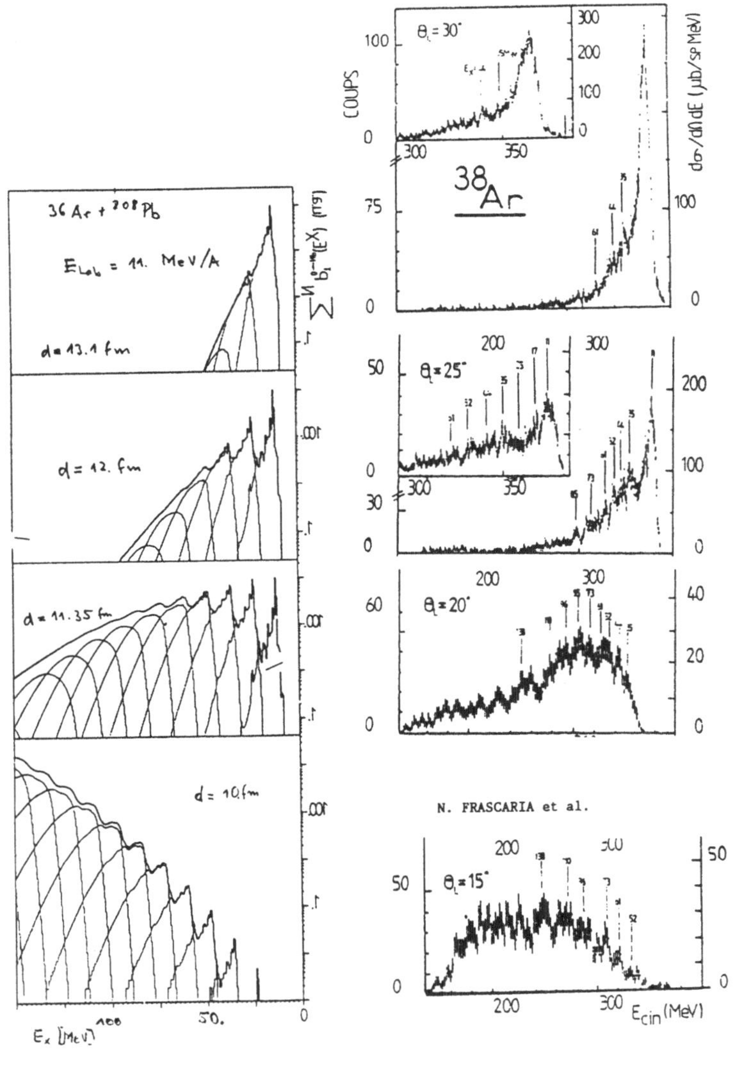

for different impact parameters d (or different corresponding deflection angles θ). We see that qualitatively the evolution and structures of the theoretical and experimental cross section are in very good agreement (the scales and units of both are not the same); theory lends support to the interpretation of the structures seen in experiment of equally spaced multiple excitation of 2^+ giant resonances.

An analogous phenomenon has been observed in inelastic atom scattering on surfaces of solids where the atom excites a whole bunch of phonons in the solid.

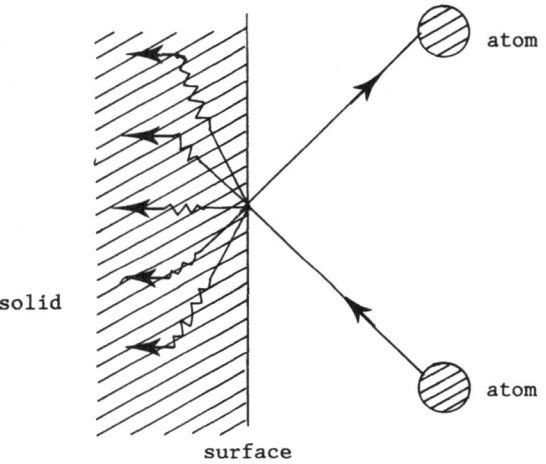

3. Vlasov equation, pseudo particles, coherent states.

A further domain of nuclear physics where coherent packets are just starting to be applied is the solution of the TDHF eqs. themselves. To this purpose rather the semiclassical limit of TDHF is considered (Vlasov equation [1,ch.13]:

$$i\hbar\dot\rho = [\ H_\rho,\rho]\ \underset{\hbar\to 0}{\to}\ \dot f(\vec R,\vec p,t) + \frac{\vec p}{m}\vec\nabla f - \frac{\partial V_\rho}{\partial \vec R}\cdot\frac{\partial f}{\partial \vec p} = 0 \tag{18}$$

In this equation $\vec R,\vec p$ are now the phase space variables of <u>real</u> space (not to be mixed up with the abstract coordinates of the Fokker Planck equation (11)). In many cases the classical limit of TDHF is justified as the following example shows [13] where the final stages of a heavy ion collision in full TDHF and

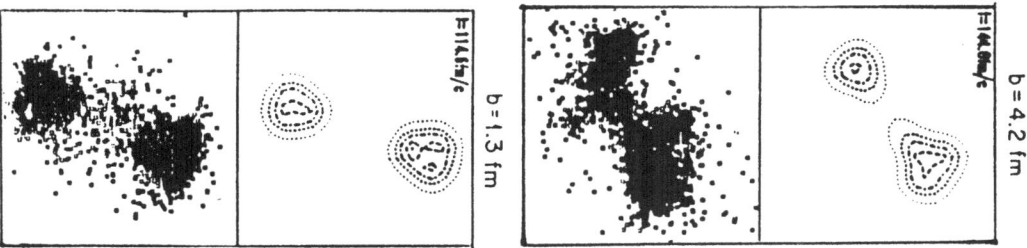

Vlasov approximation are compared.

The solution of the Vlasov equation can be formally obtained from the classical trajectories $\vec{R}_c(\vec{R}_0\vec{P}_0,t)$ and $\vec{P}_c(\vec{R}_0\vec{P}_0,t)$ where $\vec{R}_0,\vec{P}_0$ are the initial values:

$$f(\vec{R},\vec{p},t) = \int d^3R_0 d^3p_0 \ \delta(\vec{R} - \vec{R}_c(\vec{R}_0\vec{P}_0,t)) *$$

$$* \ \delta(\vec{p}-\vec{p}_c(\vec{R}_0,\vec{P}_0,t)) f_0(\vec{R}_0,\vec{P}_0,t_0) \tag{19}$$

where f_0 represents the initial phase space distribution. If we represent the δ-functions by Gaussians of finite (constant) width then (19) is represented by an (approximate) decomposition into coherent states (pseudo particles)

$$f \simeq \sum_{ij} \Delta R \ \Delta p \ f_0(\vec{R}_i\vec{P}_i,t_0) \ e^{-(\vec{R}-\vec{R}_i(t))^2/\Delta R^2} e^{-(\vec{p}-p_j(t))^2/\Delta p^2} \tag{20}$$

In principle we could make the width time dependent but "frozen" Gaussians turn out to be more effective [14]. The interest of this procedure lies in the fact that for high energy processes where many particles pass into the continuum a direct solution of the TDHF equation seems much more difficult. Also the inclusion of two body collision seems to be easier in this formalism [15]. Preliminary calculations in this direction have been undertaken by Gregoire, Remaud, Sebille, Vinet and first results obtained in the one dimensional geometry of slab collisions. A calculation of two colliding slabs by Sebille and Remaud [16] is shown in the figure.

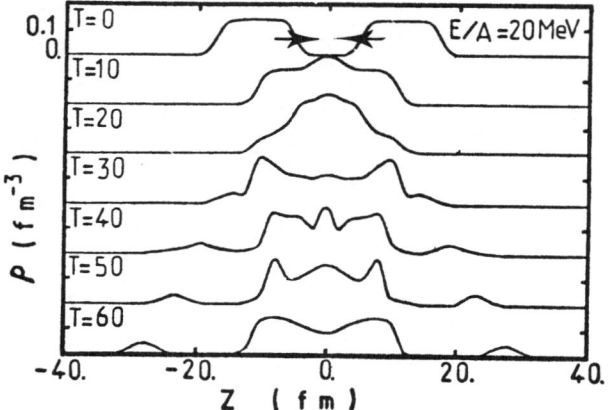

More realistic calculation in 3D for ion-ion collisions are under way.   An example of a heavy ion collision (^{12}C - ^{12}C) of high energy (84 MeV per nucleon) in 3D using a slightly different technique than the one described above (point pseudo-particles instead of coherent states) to solve Vlasov but including two body collisions is shown in the figure.

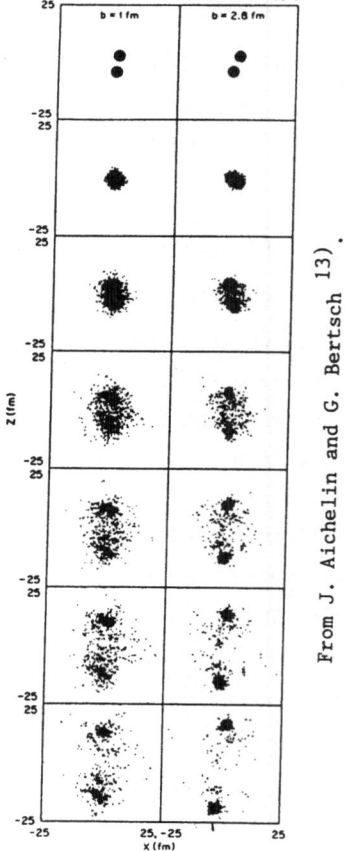

From J. Aichelin and G. Bertsch[13].

4. Giant resonances.

As was mentioned earlier, the giant quadrupole resonance plays an important role in nuclear physics. There exists a quite realistic model for this mode which is analytically solvable, and which, in phase space, leads to some sort of coherent state [17].

Using a separable residual interaction χ Q.Q where $Q = x^2+y^2-2z^2$ is the quadrupole operator and χ the strength parameter we obtain for the time dependent mean field hamiltonian:

$$h_\rho(t) = h_0 + \chi \, q(t) \, Q$$

$$q(t) = \int d^3r \, Q \, \rho(\vec{r},t) \tag{21}$$

$$h_0 = \frac{p^2}{2m} + \frac{m}{2} \omega_0^2 \, r^2 \quad .$$

Because of the quadratic coordinate dependence of $h_\rho(t)$, the Vlasov equation (18) is exact in this case and it can be solved analytically (up to the solution of classical equations of motion):

$$f(\vec{R},\vec{p},t) = \theta[\varepsilon_F - \frac{1}{2m} \sum_i \xi_i^2(t) \, (p_i - m \frac{\dot{\xi}_i}{\xi_i} u_i(\vec{R}))^2 - \frac{m}{2} \omega_0^2 \sum_i \frac{R_i^2}{\xi_i^2(t)}] \tag{22}$$

with θ the unit step function and

$$\vec{u}(\vec{R}) = \frac{1}{2} \vec{\nabla} \, Q$$

the ξ's obey the following classical equations of motion:

$$\ddot{\xi}_i - \frac{\omega_0^2}{\xi_i^3} + \Omega_i^2(t) \, \xi_i = 0 \quad ; \quad \xi_i(0) = 1 \quad ; \quad \dot{\xi}_i(0) = 0 \tag{23}$$

where the $\Omega_i(t)$ are the effective frequencies in $h_\rho(t)$ of (21). It should be noted that in (22) $\xi_x = \xi_y \neq \xi_z$. One can solve (22) directly in the small amplitude limit with the result for the giant resonance frequency

$$\omega_{2^+} = \sqrt{2} \, \omega_0 \simeq 60A^{-1/3} \text{ MeV} \tag{24}$$

which is in very good agreement with experiment. Besides the fact that (22)
represents some sort of coherent state we should notice that it represents
a non-equilibrium solution in the sense that the (local momentum distribution
is nonisotropic. As a matter of fact the momentum distribution is out of
phase with the shape deformation (see graph) a feature similar to zero sound
modes, found in ^{III}He.

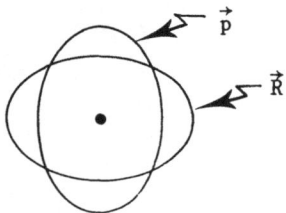

Instead of demonstrating the small amplitude solution explicitly on (22) let
us again proceed with the pseudo particle method.
In the small amplitude limit we have

$$f = f_0 + f_1 \tag{25}$$

where f_0 is the equilibrium distribution

$$f_0(\vec{R},\vec{p}) = \theta(\varepsilon_F - \frac{p^2}{2m} - \frac{m}{2} \omega_0^2 R^2) \quad , \tag{26}$$

Linearising the Vlasov equation we obtain

$$\dot{f}_1 + \frac{\vec{p}}{m} \vec{\nabla} f_1 - \vec{\nabla} V_0 \frac{\partial f_1}{\partial \vec{p}} = \vec{\nabla} V_1 \frac{\partial f_0}{\partial \vec{p}} \quad . \tag{27}$$

with an obvious meaning of V_0 and V_1. In order to solve eq. (27) we make use
of the Green's function technique and construct

$$[\frac{\partial}{\partial t} + \frac{\vec{p}\vec{\nabla}}{m} - \vec{\nabla} V_0 \frac{\partial}{\partial \vec{p}}] G^{(t-t')}(\vec{R},\vec{p},\vec{R}',\vec{p}') = \delta(\vec{R}-\vec{R}')\delta(\vec{p}-\vec{p}')\delta(t-t') \quad , \tag{28}$$

The solution of (28) is given by

$$G^{(t-t')}(\vec{R},\vec{p},\vec{R}'\vec{p}') = \delta(\vec{R}_c(\vec{R},\vec{p},t'-t)-\vec{R}')\delta(p_c(\vec{R},\vec{p},t'-t) - \vec{p}')\theta(t-t') \quad . \quad (29)$$

With

$$R_{c_i}(t'-t) = R_i \cos \omega_0(t'-t) + \frac{p_i}{m\omega_0} \sin \omega_0 (t'-t)$$

$$(30)$$

$$p_{c_i}(t'-t) = -m\omega_0 R_i \sin \omega_0 (t'-t) + p_i \cos \omega_0 (t'-t) \quad .$$

We therefore find from eq. (27):

$$f_i = i \cdot \delta(\varepsilon_F - h_0)$$

$$i(\vec{R},\vec{p},t) = i_h + \int_{-\infty}^{+\infty} dt' \int d^3R'd^3p' \; G^{(t-t')}(\vec{R},\vec{p},\vec{R}'\vec{p}') \; \vec{\nabla}V_1(\vec{R}')\cdot\frac{\vec{p}'}{m} \quad . \quad (31)$$

With i_h a solution of the homogeneous equation and

$$V_1 = x \; q_1(t) \; (R_x^2 + R_y^2 - 2R_z^2)$$

we obtain (we omit i_h in the future since it does not contribute at resonance)

$$i = \frac{2x}{m} \int_{-\infty}^{t} dt' \; q_1(t') \; [R_{cx}(R_x,p_x,t'-t) \; p_{cx}(R_x,p_x,t'-t) + R_{cy}\,p_{cy} - 2R_{cz}\,p_{cz}] .(32)$$

With (30) we then get:

$$i(\vec{R},\vec{p},t) = \frac{2x}{m} \int_{-\infty}^{t} dt' \; q_1(t') \; [\frac{1}{\omega_0} \; (\frac{p_x^2+p_y^2-2p_z^2}{2m} - \frac{m}{2} \omega_0^2(R_x^2+R_y^2-2R_z^2)) \sin 2\omega_0(t'-t)$$

$$+ (R_x p_x + R_y p_y - 2R_z p_z) \cos 2\omega_0(t'-t)] \quad . \quad (33)$$

Since we are interested only in periodic solutions we can write for $q_1(t)$:

$$q_1(t) = q_1 \; e^{i(\omega-i\eta)t} \quad\quad\quad (34)$$

and then obtain

$$i_\omega(\vec{R},\vec{p},t) = \frac{2\chi q_1(t)}{m\omega_0} \left[\frac{p_x^2+p_y^2-2p_t^2}{2m} - \frac{m}{2}\omega_0^2 (R_x^2+R_y^2-2R_z^2) \frac{2\omega_0}{\omega^2-4\omega_0^2} \right.$$

$$\left. - i \frac{2\chi}{m} (R_x p_x + R_y p_y - 2R_z p_z) \frac{\omega}{\omega^2-4\omega_0^2} \right] . \qquad (35)$$

Here q_1 is given by the following expression

$$q_1(t) = - \int \frac{d^3R\,d^3p}{(2\pi\hbar)^3} \delta(\varepsilon_F-h_0)\, i_\omega(\vec{R},\vec{p},t)(R_i^2+R_y^2-2R_z^2) . \qquad (36)$$

Using (35) in (36) yields an equation for ω

$$1 = 2\chi \frac{\omega_0^2}{\omega^2-4\omega^2} \frac{\partial}{\partial\varepsilon_F} \int d^3R\,\rho_{TF}(R)\,(R_x^2+R_y^2-2R_z^2)^2 \qquad (37)$$

$$\rho_{TF}(R) = \int \frac{d^3p}{(2\pi\hbar)^3} \theta(\varepsilon_F-h_0) .$$

Equation (37) can be transformed to

$$m = - \frac{2\chi}{\omega^2-4\omega_0^2} \int d^3R\,\frac{1}{R}\,Q^2(\vec{R})\,\frac{\partial\rho_{TF}}{\partial R} .$$

With

$$\int d^3R\,\frac{1}{R}\,Q^2\,\frac{\partial\rho}{\partial R} = -4 \int d^3R\,R^2\rho \equiv -12C$$

and

$$\chi = - \frac{m\omega_0^2}{12C} .$$

(This can be found from the usual selfconsistency condition, see Bohr/Mottelson vol. II) we finally obtain from (38):

$$\omega^2 - 4\omega_0^2 = - \frac{m\omega_0^2}{12C} \frac{2}{m} (-12C)$$

or

$$\omega_{2^+} = \sqrt{2}\ \omega_0$$

as stated above.

We have been rather explicit in the derivation here because of the following reasons: first the derivation using pseudo particles and Green's function technique is novel. Second the fact that we can restrict our pseudo particles to the $\delta(\varepsilon_F - h_0)$- shell in phase space (see (31,36)) is interesting since it limits very much the phase space volume to be considered. This would help in the numerical solution if one wanted to represent the pseudo particles by Gaussian packets in a more realistic calculation using a semiclassical approach for nuclear collective states.

Of course there are many more of such states in nuclei than the 2^+ state just discussed (such as monopole- or octupole vibrations) but we do not want to go into further details here.

(This last paragraph constitutes part of my personal research on the subject (to be published) which has been done in collaboration with H. Kohl and S. Stringari; discussions with Ch. Gregoire, G. Remaud, C. Sebille and L. Vinet are also acknowledged.)

5. Conclusions

This overview of wavepacket dynamics and the use of wavepackets and coherent states in nuclei was necessarily very brief, many items have certainly been left out and the device of subjects has surely been biased by very own preoccupations. Many more examples will however, be presented by my nuclear physics colleagues. Nevertheless I hope to have given a fair impression of the richness of time dependent processes we encounter in nuclear physics. I also hope to have pointed out the fields where wavepacket dynamics has progressed already quite a bit whereas other aspects of nuclear physics are just at the beginning and are waiting to be exploited by a wavepacket approach.

References

1) P. Ring, P. Schuck, The nuclear many body problem, Springer Verlag 1980

2) R.Y. Cusson, J.A. Maruhn, H.W. Meldner, Phys. Rev. C18 (1978) 2589

3) M. Brack et al., Rev. Mod. Phys. 44 (1972) 320

4) P. Van Leuven, this meeting.

5) T. Troudet, S. Koonin, Ann. of Phys. 154 (1984) 421

6) S. Levit, J.W. Negele, Z. Paltiel, Phys. Rev. C22 (1980) 1979

7) H. Reinhardt, Nucl. Phys. A367 (1981) 269

8) H. Hofmann, P.J. Siemens, Nucl. Phys. A275 (1977) 464

9) P. Frobrich, B. Strack, M. Duraud, Nucl. Phys. A406 (1983) 557

10) Ph. Chomaz et al., Z. Physik A318 (1984) 41

11) Ph. Chomaz, D. Vautherin, Phys. Lett. 139B (1984) 244;
    Ph. Chomaz, thèse, Orsay 1983

12) R.A. Broglia et al., Phys. Lett. 89B (1979) 22

13) J. Aichelin, G. Bertsch, Phys. Rev. 31 (1985) 1730

14) E.J. Heller, J. Chem. Phys. 75 (1981) 2923

15) Ch. Grégoire, private communication.

16) F. Sebille, B. Remaud, to be published.

17) P. Schuck, Lectures on the Random Phase Approximation, Triëste, Centre for
    Theoretical Physics, February 1984.

SEMICLASSICAL MANY-PARTICLE DYNAMICS
WITH GAUSSIAN WAVE PACKETS

K. Singer

Department of Chemistry, Royal Holloway and Bedford New College
Egham, Surrey TW20 OEX, England

W. Smith

S.E.R.C. Laboratory, Daresbury, Warrington WA44AD, England

Heller's treatment of the dynamics of Gaussian wave packets (GWPs) by the DFM variational method [1] has been adapted to perform semiclassical molecular dynamics simulations of liquids [2].

The DFM principle consists of the minimisation of the integral

$$\int |H\psi - i\hbar\Theta| d\tau \quad , \quad \Theta \equiv \frac{\partial\psi}{\partial t} \tag{1}$$

with respect to the time dependent parameters in ψ. To obtain a N-particle wave function which permits separation into N quasi one-particle Hamiltonians, the Hartree product ansatz is used

$$\psi = \prod_1^N \varphi_j \quad , \quad H = \sum_1^N (\bar{H}_j - \tfrac{1}{2}\bar{V}_j) \tag{2}$$

$$\bar{H}_j = -\frac{\hbar^2}{2m} \nabla_j^2 + \bar{V}_j \tag{3}$$

$$\bar{V}_j = \sum_{\ell \neq j} \int V_{j\ell}(|r_{j\ell}|) \varphi_\ell^*(r_\ell) \varphi_\ell(r_\ell) d^3 r_\ell \quad . \tag{4}$$

The one-particle WP has the form [1]

$$\varphi_j = \exp \frac{i}{\hbar} Q_j$$

$$Q_j = (r_j - R_j(t)) \cdot A_j(t) \cdot (r_j - R_j(t)) + P_j(t) \cdot (r_j - R_j(t)) + D_j(t) \tag{5}$$

and the time dependent parameters are $\underline{A}$, D, $\underline{P}$ and $\underline{R}$.

The minimisation leads to equations which are linear in the time derivatives of the parameters, and which can be solved numerically. The preliminary note [2] showed that the spreading of the WPs was not a serious problem and that the conservation of energy was fair. The singularities which prevent the use of most common pair potentials in (4) are avoided by the substitution of linear combinations of Gaussians

$$V_G(r) = \sum_{n=1}^{m} C_n \exp(-d_n r^2) \qquad m=2 \text{ or } 3 \qquad (6)$$

which closely fit the desired function over the required range of r. Gaussian pair potentials have the additional advantage that all integrals are readily evaluated in closed form.

Equilibrium properties of the system are computed as time averages of the relevant expectation values.

Although the results obtained for liquid and solid neon are less than satisfactory, significant improvements have resulted from three factors:

1) The use of 'semi-frozen' GWPs in which the A-parameters are constrained to be equal for all particles, but change with time, leads to a better conservation of energy, better agreement with experiment, and a substantial saving in computing time.

2) Allowance is made for the fact that for a given pair potential the interaction between two point sites is not equal to the interaction between two Gaussian distributions of sites. Formal equivalence can be obtained by adjustment of the parameters according to the mean width of the WPs in the simulation (method A). Alternatively, the parameters can be scaled empirically so as to optimise the agreement with experiment (method B).

3) By insertion of a 'classical' particle which acts as temperature probe it has been established that the temperature of the microcanonical ensemble of GWPs corresponds to the kinetic energy of the centres of the WPs.

Results

The WPs show no sign of spreading during simulation runs of up to 100 ps. This applies even to gaseous neon for V_m below 1000 ml. The rms variance of the total energy is typically 0.6% and there is no drift. The fluctuations of various properties during a typical run are illustrated in Fig. 1. Table 1 compares thermodynamic properties of neon with those based on GWP simulations with the potentials A and B, and also with those of the classical L-J fluid. With WA the energies are too high and the pressures too low; they are, however better than the results obtained with the classical L-J liquid. The WB simulations give satisfactory energies and smaller discrepancies in pressure.
The radial distribution functions (RDF) obtained from experiment, classical L-J and WBL2 simulations are compared in Fig. 3. It is seen that the WP system exagerates the quantum mechanical broadening (with the potential A this effect is even worse). Figs. 2 and 4 illustrate the difference between the centre-centre and WP-WP RDFs for liquid and solid neon near the triple point.

The velocity autocorrelation functions do not differ qualitatively from those of the classical system. The diffusion constants calculated by integration of the c.o.m. autocorrelation functions are listed in Table 2 and compared with those of the classical L-J liquid and with experimental data. The agreement -except for WBL1- is reasonably good.

We believe that the principal deficiency of the method stems from the inadequacy of the Gaussian function as approximate solution of the Schrödinger equation for particles moving in a harshly repulsive potential field. The diffuseness of the wings can only be reduced by a narrowing and enhancement of the peak, which entails an increase in kinetic energy. As the entries in table 1 show, the ratio of quantum mechanical to c.o.m. kinetic energies in the WP simulations is roughly 1, whereas calculations based on the Wigner expansion [3] and on neutron scattering data [4] indicate that it should be 0.5. The deficiency of the GWPs can partly be masked by potential modelling. Significant progress, however, is unlikely without recourse to more sophisticated trial functions.

The potential functions used in our WP simulations are given in table 3.

References

1) E.J. Heller, J. Chem. Phys. 64 (1976) 63
2) N. Corbin and K. Singer, Molec. Phys. 46 (1982) 671
3) J-P. Hansen and J.J. Weis, Phys. Rev. 188 (1969) 314
4) V.F. Sears, Can. J. Phys. 59 (1981) 555

	V_m ml/mol	$\langle T \rangle$ K	$\langle PE \rangle$ kJ/mol	$\langle KE \rangle$ kJ/mol	$\langle E \rangle$ kJ/mol	$\langle P \rangle$ Mpa	$\langle x^2 \rangle$ $10^{-22}m^2$
WAL1	16.20	25.2	-1.79	0.64	-1.15	-10.8	2.2
CLJ	"	"	-1.80	0.31	-1.49	-13.5	-
FXP	"	"	-	-	-1.26	1.7	-
WAL2	18.0	34.3	-1.56	0.74	-0.82	-0.2	2.4
CLJ	-	"	-1.57	0.43	-1.14	-1.4	-
EXP	-	"	-	-	-0.97	6.6	-
WAS1	13.0	20.8	-2.35	0.65	-1.69	2.1	2.0
EXP	13.7	"	-	-	-1.76	0.0	-
WAG1	721.0	28.6	-0.09	0.42	0.33	0.3	13.4
CLJ	"	"	-0.06	0.36	0.29	0.3	-
WUAL1	16.16	24.8	-1.81	0.74	-1.06	-29.8	3.2
CLJ	"	"	-1.80	0.31	-1.49	-13.2	-
EXP	"	"	-	-	-1.28	0.9	-
WBL1	16.16	24.8	-1.98	0.67	-1.31	-3.6	2.1
CLJ	"	"	-1.80	0.31	-1.49	-13.2	-
EXP	"	"	-	-	-1.28	0.9	-
WBL2	18.0	33.0	-1.72	0.75	-0.98	-0.2	2.2
CLJ	"	"	-1.59	0.41	-1.18	-4.4	-
EXP	"	"	-	-	-0.99	3.8	-
WBS1	13.7	20.1	-2.45	0.66	-1.79	6.2	1.8
EXP	"	"	-	-	-1.78	0.0	-

WA.., WB..: Wp-simulations with potentials determined according to method A and B respectively L = Liquid, S = Solid, G = Gas.

Columns 4-9: potential energy, kinetic energy, total energy, pressure, WP-width. The pair potentials in the "A"-simulations correspond to these widths.

WUAL1 is a simulation in which the WPs were not constrained to equal widths.

Table 1

	V_m	T	D_{WP}	D_{CLJ}	D_{EXP}
	ml/mol	K	$10^{-9} m^2/sec$	$10^{-9} m^2/sec$	$10^{-9} m^2/sec$
WAL1	16.20	25.2	1.16	1.31	0.93 ± 0.1
WAL2	18.0	34.3	2.60	3.01	2.77 ± 0.2
WAG1	721.0	28.6	182	216	-
WBL1	16.16	24.8	0.62	0.86	0.83 ± 0.1
WBL2	18.0	33.0	2.08	2.64	2.35 ± 0.2

Table 2: Selfdiffusion Constants.

1. Approximation to the Lennard-Jones potential for neon,

$$V(x) = 4\varepsilon(x^{12}-x^6) \ , \ x = r/\sigma \ , \quad = 2.789.10^{-10}m \ , \quad \varepsilon = 0.508.10^{-21}J$$

$$V_G(r) = \sum_1^3 C_n \exp(-d_n r^2)$$

	C_1	C_2	C_3	d_1	d_2	d_3
	\multicolumn{3}{c}{$/10^{-20}J$}	\multicolumn{3}{c}{$/10^{20}m^{-2}$}
site-site (width = 0)	450.1	-0.3653	-0.01427	1.0993	0.1985	0.0585

A. adjusted for width according to (26):

m.sq. width $\langle x^2 \rangle$ $/10^{-22}m^2$	C_1	C_2	C_3	d_1	d_2	d_3
	\multicolumn{3}{c}{$/10^{-20}J$}	\multicolumn{3}{c}{$/10^{20}m^2$}
2.0	516.75	-0.3742	-0.01437	1.2053	0.2017	0.05877
2.2	524.31	-0.3751	-0.01438	1.2170	0.2020	0.05880
2.4	532.06	-0.3760	-0.01439	1.2290	0.20236	0.05883

B. empirical adjustment:

	537.1	-0.4075	-0.0157	1.1315	0.1954	0.0570

this potential is obtained by a width correction for $\langle x^2 \rangle = 1.2.10^{-22}m^2$, followed by multiplication of the C_n by 1.1 and of the d_n by 0.975. It was used for different state points without further correction for width.

Table 3: Three-term Gaussian pair potentials.

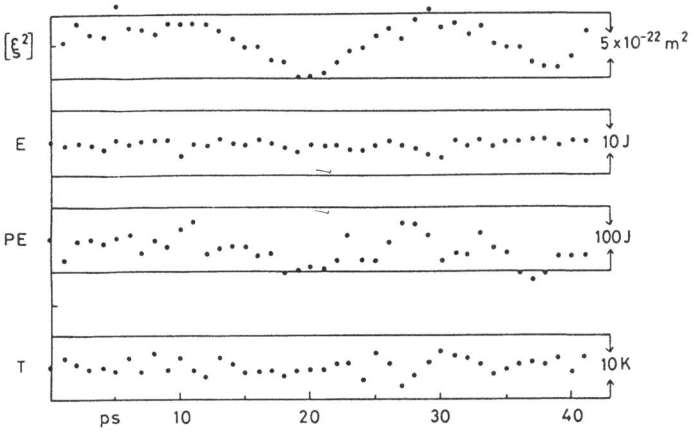

Figure 1: Fluctuations in a typical run.

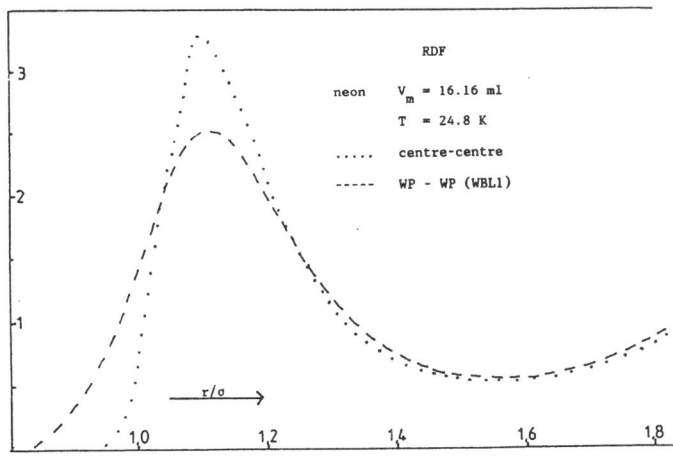

Figure 2.

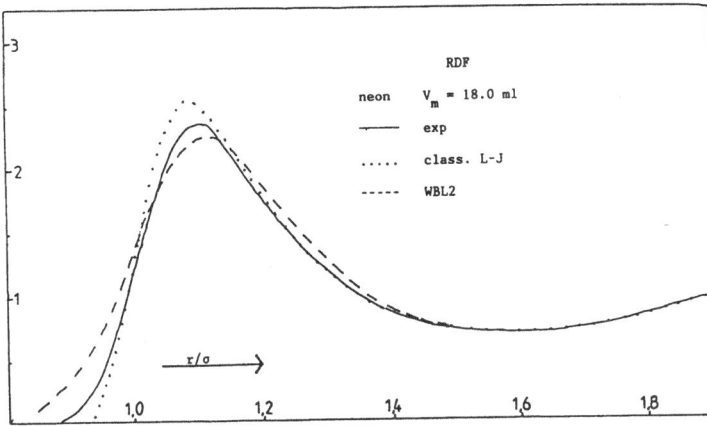

Figure 3.

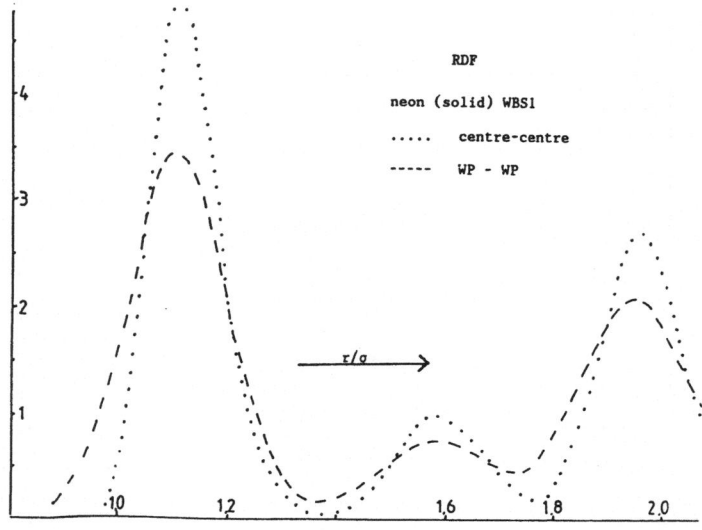

Figure 4.

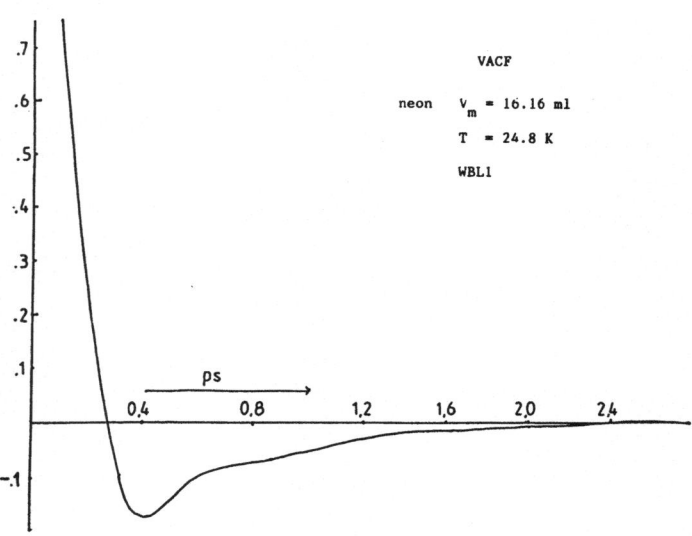

Figure. 5

GAUSSIAN WAVE PACKET DYNAMICS EXPRESSED IN THE CLASSICAL INTERACTION PICTURE

R.T. Skodje

Department of Chemistry, University of Colorado, Campus Box 215
Boulder, CO 80309, USA

Many semiclassical methods imbed exact classical trajectories in models of quantum phenomena. The purpose of the present work [1,2] is to demonstrate the utility of representing the classical mechanics in the classical interaction picture for one such method, viz. the time-dependent Gaussian wave packet method (GWPM). The classical interaction picture (GIP) is invoked by applying a canonical transformation which is entirely analogous to the unitary transformation of the quantum interaction picture. If we split the classical Hamiltonian into two terms

$$H_0^c = H_0^c + H_I^c \tag{1}$$

then the basic strategy of the GIP is to transform to new coordinates in which the motion generated by H_0^c is in a sense, subtracted off. In the GWPM, a time-dependent wave function is represented by a superposition of Gaussian packets whose centers follow classical trajectories [3,4]. Errors in the time evolution of the wave function are caused by anharmonicities in the potential function. To show how use of the CIP can minimize certain of these errors, consider the example of a collinear collision of an atom with an anharmonic oscillator. The initial wave function, representing a particular incoming oscillator state and a distribution peaked at correct translational momentum, is accurately found by diagonalizing the asymptotic Hamiltonian in the Gaussian basis. As this state time evolves in the asymptotic region, its accuracy degrades as a result of anharmonicities of the oscillator potential. Skodje and Truhar [5] have shown that large inaccuracies in transition probabilities can result from this degradation especially for low-energy collisions which require long times to establish the asymptotic conditions. By using the CIP the asymptotic motion can be subtracted off leading to trajectories which are nearly constant in time in the asymptotic regions and strongly time-dependent only during a brief interval surrounding the main part of the collision. This leads to a GWPM wave function which is constant in the asymptotic region. Thus the wave function will remain a good representation of the incoming state until the interaction potential becomes important and a major source of error is removed from the method.

Inelastic scattering is but one example of how the CIP can be profitably used in the GWPM. The central idea of the procedure is to transform away the time-dependence of the wave function due to the zeroth order quantum Hamiltonian H_0^Q. In general whenever H_I^Q is small for all or part of the time domain and H_0^Q is easily diagonalizable but anharmonic, then the semiclassical time propagation of the wave packets should be accurate for much longer times using the CIP. The reason is that we avoid the inaccuracies of long time propagation in the anharmonic H_0^Q potential. Since the H_0^Q generated motion is not very interesting (since we gain nothing beyond the initial time-dependent diagonalization), the CIP method has the desired property of focusing on the dynamics generated by H_I^Q.

We now briefly review the formal derivation of the CIP [1]. The basic idea is to subtract off the motion generated by H_0^c. This can be accomplished by propagating a trajectory forward from time 0 to time t using the full Hamiltonian H^c, then propagating backwards from t to 0 using the Hamiltonian H_0^c. The time-dependence of the resultant trajectory comes from the difference between H^c and H_0^c, viz. H_I^c. To formulate the CIP, we define three sets of dynamical variables: $(\vec{q},\vec{p}),(\vec{q}_0,\vec{p}_0)$, and $(\vec{q}_I,\vec{p}_I)$. The set $(\vec{q},\vec{p})$ is defined as the coordinates propagated by H^c; $(\vec{q}_0,\vec{p}_0)$ are the coordinates propagated by H_0^c; and $(\vec{q}_I,\vec{p}_I)$ are the CIP coordinates. We express the time-dependence of these variables, using $\vec{p}_0$ as an example, as

$$\vec{p}_0 = \vec{p}_0(\vec{X},\vec{Y};t_0,t) \quad .$$
(2)

Here $\vec{p}_0$ is the result of propagating the initial point $(\vec{X},\vec{Y})$ at time t_0 to a final time t. If $t_0 > t$ a backward time propagation is indicated. Our physical interpretation of the CIP suggests we write

$$\vec{q}_I = \vec{q}_0(\vec{q},\vec{p};t,0)$$
(3a)

$$\vec{p}_I = p_0(\vec{q},\vec{p};t,0) \quad .$$
(3b)

It is easy to show that $(\vec{q}_I, \vec{p}_I)$ must satisfy, in component language,

$$\dot{q}_{iI} = \frac{\partial H_I^c}{\partial p_{iI}} \tag{4a}$$

$$\dot{p}_{iI} = - \frac{\partial H_I^c}{\partial q_{iI}} \tag{4b}$$

The CIP is most useful when the transformations eqs. (3) are analytically known or when a good analytic approximation to them exists.

In the semiclassical GWPM a time-dependent wave function is written as

$$|\psi(t)\rangle = \sum_i c_i |g_i(t)\rangle \tag{5}$$

where, in Cartesian coordinates, we have

$$\langle \vec{R}|g_i(t)\rangle = \exp[\frac{i}{\hbar} \{(\vec{R}-\vec{R}(t)).\hat{A}(t).(\vec{R}-\vec{R}(t)) + \vec{P}(t).(\vec{R}-\vec{R}(t)) + \gamma(t)\}] \tag{6}$$

where $(\vec{R}(t), \vec{P}(t))$ is a classical trajectory which may be different for each Gaussian. One strategy is to keep the coefficients $\{c_i\}$ constant in time and choose the time-dependent Gaussian parameters so that each separate packet well approximates a solution to the time-dependent Schrödinger equations[3-5]. If the total potential is expanded to second order in a Taylor series about the center of Gaussian, then a simple set of differential equations for the Gaussian parameters are obtained.

We now transform the GWPM to the CIP. The Schrödinger equation in the interaction picture is

$$H_{I,eff}^Q |\psi(t)\rangle_I = i\hbar \frac{\partial}{\partial t} |\psi(t)\rangle_I \tag{7}$$

with

$$|\psi(t)\rangle_I = \sum_k c_k |g_k(t)\rangle_I \tag{8}$$

and

$$\langle \vec{R}|g(t)\rangle_I = \exp[\,\frac{i}{h}\,\{(\vec{R}-\vec{R}_I(t)).\hat{A}^I(t).(\vec{R}-\vec{R}_I(t))$$

$$+\ \vec{P}_I(t).(\vec{R}-\vec{R}_I(t))+\gamma^I(t)\}]. \tag{9}$$

The trajectory parameters $(R_I(t),\ P_I(t))$ solve eq. (4) and thus are time propagated by the interaction Hamiltonian H_I^C.

The effective Hamiltonian is obtained by the following prescription.

$$H_{I,eff}^Q = H_I^C(t) + \sum_k \ (\ \frac{\partial H_I^C}{\partial R_{kI}}\ (R_k - R_{kI}(t)) + \frac{\partial H_I^C}{\partial P_{kI}}\ (P_k - P_{kI}(t)))$$

$$+\ \frac{1}{2}\sum_{jk}\ (\frac{\partial^2 H_I^C}{\partial R_{jI}\,R_{kI}}\ (R_j - R_{jI}(t))(R_k - R_{kI}(t))$$

$$+\ \frac{1}{2}\ \frac{\partial^2 H_I^C}{\partial P_{jI}\,P_{kI}}\ (P_j - P_{jI}(t))(P_k - P_{kI}(t))$$

$$+\ \frac{\partial^2 H_I^C}{\partial R_{jI}\,P_{kI}}\ (P_j - P_{jI}(t))(P_k - P_{kI}(t)) + (P_k - P_{kI}(t))(R_j - R_{jI}(t))]) \tag{10}$$

where the derivatives of H_I^C are evaluated at $(\vec{R}_I(t),\vec{P}_I(t))$. It is easily verified that each packet will individually solve the effective Schrödinger equation when $(\vec{R}_I(t),\vec{P}_I(t))$ solves eq. (4) and when

$$\dot{A}_{ij}^I(t) = -2\sum_{k\ell} A_{ik}^I(t)\ \frac{\partial^2 H_I^C}{\partial P_{kI}\partial P_{\ell I}}\ A_{\ell j}^I(t) - \frac{1}{2}\ \frac{\partial^2 H_I^C}{\partial R_{iI}\partial R_{jI}}$$

$$+\sum_k (\frac{\partial^2 H_I^C}{\partial P_{kI}\,R_{iI}}\ Q_{kj}(t) + A_{ik}^I(t)\ \frac{\partial^2 H_I^C}{\partial P_{kI}\partial R_{jI}}) \tag{11}$$

$$\dot{\gamma}^I(t) = ih\sum_{ij} \frac{\partial^2 H_I^C}{P_{iI}\,P_{jI}}\ A_{jI}^I(t) + \frac{ih}{2}\sum_i \frac{\partial^2 H_I^C}{P_{iI}\,R_{iI}} + \vec{R}_I(t).\vec{P}_I(t) - H_I^C(t). \tag{12}$$

In the asymptotic regions where H_I^C becomes vanishingly small, it is seen that all the packet parameters become constant. Thus the total wave function will not degrade as the result of free motion. We must note that the errors induced by the anharmonicity of the oscillator potential have not been eliminated. The anharmonicity has been shifted to H_I^C which operates for a much shorter period of time than does H_0^C.

In a numerical test, the method described was applied to the Clark-Dickinson problem of the collinear collision of an atom with a Morse oscillator diatom [6]. It was shown [2] that a substantial portion of the error incurred by the usual GWPM could be eliminated by transforming to the CIP.

References

1) R.T. Skodje, Chem. Phys. Lett. 109 (1984) 221.

2) R.T. Skodje, Chem. Phys. Lett. 109 (1984) 227.

3) E.J. Heller, J. Chem. Phys. 62 (1975) 1544.

4) E.J. Heller, J. Chem. Phys. 65 (1976) 4979.

5) R.T. Skodje and D.G. Truhlar, J. Chem. Phys. 80 (1984) .

6) A.P. Clark and A.S. Dickinson, J. Phys. B6 (1973) 164.

GENERAL ASPECTS
OF WAVE PACKET DYNAMICS

P. Van Leuven

Dienst Teoretische en Wiskundige Natuurkunde
University of Antwerp, R.U.C.A.,
Groenenborgerlaan 171, 2020 Antwerp, Belgium.

0. Introduction

In this introduction we enumerate a number of general aspects of wave packets and their use. Although they may occur in a different formal language in different fields of research, the statements which follow are well-known. In view of the diversified specialisations represented at this meeting, they might serve as a framework in which to classify concepts and methods.

1. Classical wave packets

In classical physics a "wave packet" (WP) is a wave of which the amplitude is small outside a domain of width Δx and inside this domain is approximately periodic with wavelength λ. It then follows that its Fourier transform has a large amplitude and a stationary phase in the neighbourhood, of width Δk, of the wave number $k=2\pi/\lambda$. It also follows that the product $\Delta x . \Delta k$ has a lower limit. The dynamics of the classical wave packet is determined by the wave equation and the dispersion relation $\omega(k)$.

2. The Gaussian wave packet

In elementary quantum mechanics of 1 particle in 1 dimension we encounter the Gaussian WP (GWP)

$$\psi(x) \sim \exp(ipx) \exp[-\frac{1}{2}(\frac{x-q}{b})^2] \qquad (\hbar=1) \qquad . \qquad (1)$$

These wave functions constitute a 3-parametric manifold $\psi(x|qpb)$. The physical meaning of the parameters is determined by

$$q = \langle\psi|x|\psi\rangle \quad , \qquad p = \langle\psi|\hat{p}|\psi\rangle \quad , \qquad \frac{b^2}{2} = \langle\psi|(x-q)^2|\psi\rangle \quad . \quad (2)$$

The GWP has a number of special properties

a) it has a minimum uncertainty product

$$\Delta x \Delta p = \frac{1}{2} \tag{3}$$

b) it can be written as a displaced oscillator ground state

$$\psi \sim \exp(ip\hat{x})\exp(-iq\hat{p})\exp[-\frac{1}{2}(\frac{x}{b})^2] \quad . \tag{4}$$

The oscillator ground state is displaced in space $(x \rightarrow x-q)$ and in momentum (boost).

c) it is a coherent superposition of oscillator states.
Putting

$$\alpha = \frac{1}{\sqrt{2}} (\frac{q}{b} + ibp) \tag{5}$$

we can write

$$\psi \sim \sum_n c_n \varphi_n \tag{6}$$

with

$$c_n = \exp(-\frac{1}{2}|\alpha|^2) \frac{\alpha^n}{\sqrt{n!}} \quad . \tag{7}$$

Typical for the coefficients c_n is that the amplitude $|c_n|$ has a maximum near $\langle\psi|\hat{n}|\psi\rangle$ and the phases are equidistant.

d) it is an eigenstate of the phonon annihilation operator

$$\hat{a}\psi = \alpha\psi \quad . \tag{8}$$

If $\psi(x)$ is considered as an initial state, its time evolution is determined by

$$\psi(xt) = \exp(-iH(t)\psi(x) \quad . \tag{9}$$

For a general Hamiltonian H, it evolves out of the manifold of GWP. For Hamiltonians with quadratic potentials it remains a GWP.

$$|\psi|^2 \sim \exp - (\frac{x-q(t)}{b(t)})^2 \quad . \tag{10}$$

In free space the center q has a uniform translation and b spreads.

$$q(t) = q + \frac{p}{m} t$$

$$b(t) = b(1+t^2/m^2b^4)^{1/2} \quad . \tag{11}$$

In an oscillator field with $\sqrt{\frac{1}{m\omega}} = b$ the GWP is coherent, i.e. the center q has a harmonic motion with frequency ω and b is constant.

$$q(t) = q \cos \omega t + \frac{p}{m\omega} \sin \omega t \tag{12}$$

$$b(t) = b \quad .$$

In an oscillator field with $\sqrt{\frac{1}{m\omega}} \neq b$ both the center q and the width b have harmonic motions with frequencies ω and 2ω repectively

$$q(t) \sim \cos \omega t$$

$$b(t) \sim \cos 2\omega t \quad . \tag{13}$$

3. Generalized wave packets

Several generalizations of the elempentary GWP, usually referred to as "coherent states", have been proposed by using one of the properties a,b,c as definition. We can distinguish.

minimum uncertainty WP (Nieto)

eigenstates of annihilation operators (Barut)

displaced ground states (Perelomov)

For these generalized WP there is no natural dynamics in which the WP behaves strictly coherent; they spread but if, well-defined, the spreading time is much larger than the classical period.

As a special example we mention the <u>Generalized Gaussian WP</u> (GGWP)

$$\psi(x) \sim \exp - (Ax^2+Bx+C) \quad . \tag{14}$$

This contains 3 complex parameters (or matrices in the multidimensional
case). The parameter C determines the norm and phase. The coefficients A
and B contain 4 real parameters which can be interpreted in terms of q,p,b
and c as follows:

$$\psi \sim \exp(ip\hat{x})\exp(-iqp)\exp(i\frac{c}{b}\,x^2)\exp[-\ell nb(\hat{x}\hat{p}+\hat{p}\hat{x})]\exp(-\frac{1}{2}x^2) \quad . \tag{15}$$

ψ may be considered as a displaced reference oscillator groundstate.
The "displacement" contains a dilation (x→x/b) as well as a translation. There
is a boost associated both with the dilation (velocity field 2cx/mb) and the
translation (velocity p/m).
The minimum uncertainty is lost

$$\Delta x \Delta p = \frac{1}{2}\,(1 + (2bc)^2)^{1/2} \quad . \tag{16}$$

Another example of Generalized WP concerns the angular momentum or
<u>rotational coherent states</u>. These can be presented as superpositions of
angular momentum eigenstates

$$\psi \sim \sum C_{jm}\,\psi_{jm} \quad . \tag{17}$$

The amplitudes $|C_{jm}|$ are peaked in the jm-plane and the phases are equidistant.
These properties assure that ψ has a large deformation (quadrupole moment) and
a well-defined orientation.

Specific expressions for C_{jm} can be derived. In the Atkins-Dobson
definition ψ is constructed as an eigenstate of the spin-annihilation operators

$$C_{jm} = \exp\,[-\frac{1}{2}(|\alpha|^2+|\beta|^2)]\,[\frac{\alpha^{j+m}}{\sqrt{(j+m)!}}\,\frac{\beta^{j-m}}{\sqrt{(j-m)!}}\,] \quad . \tag{18}$$

ψ depends on 4 real parameters which can be related to $\langle \hat{j}_x \rangle$, $\langle \hat{j}_y \rangle$, $\langle \hat{j}_z \rangle$ and an overall phase.

Another possibility is a definition through a displaced reference state where the displacement involves besides rotation (angular momentum) operators also quadrupole operators

$$\psi \sim \exp \, i[\sum_i (\omega_i \hat{j}_i) + \sum_\mu q_\mu \hat{Q}_\mu]\psi_0 \qquad . \tag{19}$$

4. Dynamical aspects

The time dependence of the WP can be studied under the exact (Schrödinger) dynamics. This is feasible for low dimensional systems (Nieto). In general the WP loses its form. In certain cases it may be an interesting approximation to consider an approximate time dependence under the restricted dynamics that the WP remains in the manifold. This means that the time dependence is put into the parameters: $\psi(x|\alpha(t))$. It is here that the semiclassical aspect of WP theory enters because the quantum principle of superposition is lost.

Various forms of resticted dynamics can be considered, e.g.: Ehrenfest equations of motion: If the WP parameters α can be written as expectation values of operators $\hat{\alpha}$ then $\dot{\alpha} = \langle \psi | -i[\hat{\alpha},H] | \psi \rangle$. If moreover the commutators $-i\langle [\alpha,H] \rangle$ can be expressed in terms of the α as $f(\alpha)$, we obtain the equations of motion $\dot{\alpha} = f(\alpha)$. These are a generalization of the Ehrenfest equations for q and p.

Time Dependent Variational Principle (TDVP): The variational principle

$$\delta \int_0^t \langle \psi | H - i \frac{\partial}{\partial t} | \psi \rangle dt = 0 \tag{20}$$

if applied to the manifold $\psi(x|\alpha(t))$, where the parameters can be grouped in conjugate pairs (q,p), leads to Hamilton's equations of motion

$$\dot{\alpha} = \{\alpha, \mathcal{H}\} \tag{21}$$

where the Poisson brackets $\{ \ \}$ and the Hamilton function $\mathcal{H}$ are derived from ψ in a straightforward way.

In the special example of the GGWP, Ehrenfest's equations and TDVP are equivalent and the equations of motion are

$$\dot{q} = \frac{p}{m} \qquad\qquad\qquad \dot{b} = 2\frac{c}{m}$$

$$\dot{p} = -\frac{\partial}{\partial q}\langle V\rangle \qquad\qquad \dot{c} = \frac{1}{2mb^3} - \frac{\partial}{\partial b}\langle V\rangle \quad .$$

$$(22)$$

For potentials V which are not harmonic, there is a coupling between b and q, i.e. between dispersion and translation. The dynamics of the wave packet is thus simulated by the motion in a 2-dimensional potential $\langle V\rangle$ which is a Gauss transform of V.

$$\langle V\rangle = \frac{1}{\sqrt{\pi}b} \int_{-\infty}^{\infty} V(x+q)\ \exp - \left(\frac{x}{b}\right)^2\ dx \quad . \qquad (23)$$

5. Phase space aspects

WP have N classical degrees of freedom built in. With these there corresponds a 2N-dimensional phase space of conjugate pairs of parameters $(q,p...)$. If the WP is prepared at t=0 by choosing initial values $(q_0 p_0 ...)$ the equations of motion trace out a trajectory on the energy hypersurface.
There are several ways of representing the motion in phase space.

Poincaré sections: For the study of the structure of phase space one uses the Poincaré plots. These are plots of the penetration of the trajectory through a 2-dimensional subspace. The pattern of penetration points reveals the periodic, quasi-periodic or chaotic nature of the trajectory.

Wigner distribution: In the case of the GGWP we have a 4-dimensional phase space. The motion can be pictured in a 2-dimensional plot if we consider the Wigner distribution $W(Q,P)$ defined as follows

$$W(Q,P) = \iint \psi^*(x)\psi(x')\ e^{iP(x-x')}\ \delta\left(\frac{x+x'}{2} - Q\right)\ dxdx' \quad . \qquad (24)$$

For the GGWP this is of the form

$$W(QP) \sim \exp -[A(Q-q)^2 + 2B(Q-q)(P-p) + C(P-p)^2] \qquad . \qquad (25)$$

The equidensity contours of W are ellipses centered about the point (q,p) with orientation and excentricity determined by (b,c). The ellipses together with the projection of the trajectory on the (q,p)-plane show the phase space behaviour of the WP. For a quadratic potential the trajectory is periodic and the orientation of the Wigner ellipse follows the tangent to the (qp)-trajectory.

6. Many particle aspects

In the realistic case of an A particle system in 3 dimensions the wave packets depend on a set of dynamical variables $\bar{r}_i$ and a set of WP parameters α_j

$$\psi(r_i \ldots r_A | \alpha_1 \ldots \alpha_n) \qquad . \qquad (26)$$

Two extreme situations may be distinguished.

Collective motion: In nuclear physics one is interested in collective motion. Here the WP is generated by the action of a collective operator $\sum_i \hat{\Omega}(\bar{r}_i)$. The simplest example is the "breathing" motion associated with

$$\psi \sim \exp[-i\ell nb \sum_{i=1}^{A} (\bar{r}_i \bar{p}_i + \bar{p}_i \bar{r}_i)]\psi_0 \sim \psi_0(\frac{r_1}{b} \ldots \frac{r_A}{b}) \qquad . \qquad (27)$$

Here the A particle problem is reduced to a single quasi-particle problem.

Individual motion: In molecular physics one is interested in the motion of the nuclei over the multi-dimensional potential hypersurface.
Here we may associate a WP to each nucleus

$$\psi \sim \prod_{i=1}^{A} \exp -\frac{1}{2} (\frac{\bar{r}_i - \bar{q}_i}{b_i})^2 \qquad . \qquad (28)$$

In this case the number of WP parameters may even be larger than the number of nuclear degrees of freedom. Still more extreme is the case (Heller) where the nuclei are fully correlated through a 3A × 3A matrix M:

$$\psi \sim \exp \sum_{i,j} (\bar{r}_i - \bar{q}_i) M_{ij} (\bar{r}_j - \bar{q}_j) \quad . \tag{28'}$$

7. Applicational aspects

WP may be used as basis functions in Hilbert space. In molecular physics e.g. they may be used to calculate the energy levels of the nuclear potential surface. In the case of 1 dimension the manifold of GWP with continuous parameter q is an overcomplete set. There exist many denumerably infinite discrete subsets which are complete. One way of discretization which might be well adapted to the Hamiltonian could be obtained by discretizing the time-parameter along the semi-classical trajectory

$$\psi_i = \psi(x | \alpha(t_i)) \quad . \tag{29}$$

Gradual sophistication of WP-basis sets may result in an increased rate of convergence:
The set of translated Gaussians $\exp -\frac{1}{2}(\frac{x-q_i}{b})^2$ may be improved by boosting ($\exp ip_i x$) which introduces nodes and may speed up convergence for highly excited energy levels. Dilation of Gaussians favours adapting the basis functions to the form of the potential (e.g. penetrate barriers). The dilational boost may further improve the phase space behaviour of the basis set.

In nuclear physics the key problem where WP can be used is the microscopic description of collective motion. Instead of considering collectivity as an output of large shell-model calculations, one uses collectivity as an input and constructs a basis of collective WP to generate a collective subspace.
The form of the WP is suggested by the type of collective motion to be studied. Whether the system can sustain this type of motion is reflected by the fact that the collective subspace is invariant under the Hamiltonian or not.
Because collective motion is usually understood in terms of classical pictures (rotation, vibration...) and as classical pictures of motion are always time-dependent, the WP dynamics is necessary to analyse the "nature" of the collective motion.

8. Group theoretical aspects

Group theoretical properties of WP and Hamiltonians may play an important role in the application of WP. We recall the GGWP. The 4 operators x, p, x^2 and $xp+px$ were used to construct the WP from the reference state. These can conveniently be represented with creation and annihilation operators: a, a^+, aa and a^+a^+. If these are supplemented by the unit operator A and the operator a^+a+aa^+ we obtain a set of 6 operators which together form the algebra of the inhomogeneous symplectic transformations I $Sp(2,\mathbb{R})$. The oscillator ground state ψ_0 can be considered as an extremal state of an irreducible representation of the I $Sp(2,\mathbb{R})$ group. It is annihilated by the operators a and aa, and it is an eigenstate of a^+a+aa^+. The group operators generated by 1 and a^+a+aa^+ acting on ψ_0 yield only a phase factor, they constitute the stability group S. Hence the manifold of GGWP can be considered as a function over the coset space of I $Sp(2,\mathbb{R})/S$ and realized by the functions

$$\psi \sim \exp(\alpha a^+ + \alpha^* a)\exp(\beta a^+ a^+ + \beta^* aa)\psi_0 \quad . \tag{30}$$

Perelomov has generalized this example to any arbitrary Lie group. He defines the coherent state by the following procedure: consider a Lie group G with elements g and an irreducible unitary representation T(g) acting in the Hilbert space $\mathcal{H}$. Choose a fixed vector ψ_0 in $\mathcal{H}$. The manifold of coherent states is then obtained by the action of T(g) on ψ_0 for all g in G/S where S is the stability group of ψ_0, i.e. the subgroup for which $T\psi_0$ and ψ_0 differ by a phase factor.

9. Quantizational aspects

The problem arises of how to get quantized solutions of the classical equations as an approximation to the quantum states of the system.

A direct quantization of the classical problem by <u>Bohr-Sommerfeld rules</u> has been applied by Arvieu.

In the case of the TDVP Kan and Griffin have introduced the <u>Gauge Invariant Periodic Quantization</u> (GIPQ) which is as follows:

a) let $\psi(x,t)$ be a solution of the time-dependent norm-changing variational
   principle
b) its gauge invariant part ψ_g is given by $\psi = \psi_g \exp(-i\langle H \rangle t)$
c) take the gauge invariant periodic solutions for which $\psi_g^P(t+T) = \psi_g^P(t)$
d) determine those $\psi_n = \psi_g^P \exp(-i\langle H \rangle t)$ for which

$$\int_0^T \langle \psi_n | i \frac{\partial}{\partial t} | \psi_n \rangle dt = 2\pi n \quad . \tag{31}$$

These ψ_n are approximations to the stationary states of H. This method
has been tested in the case where the manifold is the full Hilbert space.
It turns out that the GIPQ yields redundant solutions which can however be
recognized by the fact that they vanish upon energy projection.
The method has recently been applied for the multidimensional and non-
separable case (Ploszajczak).

10. Stochastic aspects

It is well-known that the classical equations with 2 or more degrees of
freedom display two kinds of orbits: "regular" (periodic or quasi-periodic)
orbits and "chaotic" orbits. As the WP are used as a bridge between quantum
physics and classical physics, the problem arises whether this distinction
persists in the quantum properties and what is the quantum criterion for chaos.
Here WP can be used as probes for chaotic behaviour and the rate of spreading
as a criterion (Moiseyev, Peres).

CONCLUSION

J. Broeckhove, L. Lathouwers and P. Van Leuven

Dienst Teoretische en Wiskundige Natuurkunde
University of Antwerp, R.U.C.A.,
Groenenborgerlaan 171, 2020 Antwerp, Belgium.

In the course of many of the talks and discussions a number of areas
of experience common to both molecular and nuclear physics could be defined.
Obviously also a number of striking differences have appeared. It is useful
to consider first the differences between the molecular and nuclear physicists'
approach to wave packet theory. It is here that a transfer of knowledge can
possibly be established.

A first striking distinction is that in nuclear physics the definition
of wave packets is taken in a more general sense than in molecular phycics.
This is not a fundamental difference; it is related to the fact that in
molecular physics, wave packets have been mainly used to describe the motion
of nuclei on the potential energy surface,whereas in nuclear physics wave
packets are adapted to various forms of motion: individual particle motion
with (time dependent) Hartree-Fock wave packets, collective vibrations and
rotations with coherent state wave packets. In molecular physics the motion
of wave packets is practically exclusively restricted to the (generalized)
Gaussian wave function with translational degrees of freedom. In the case of nuclear
fission and heavy ion reactions nuclear physics comes close to the molecular
description.

A second major distinction concerns the relation to classical physics.
In molecular problems the physical situation is usually such that a purely
classical description is meaningful. Hence wave packet theory tends to be
considered as an improvement on the classical results. Although this
improvement is a drastic one conceptually it may be only a minor one in terms
of numerical results. In nuclear physics we also encounter the situation
where one is far from classical physics and one must stay within the quantal
description. In this case (e.g. low lying collective states) wave packet
theory is regarded as an approximation to the complete quantum theory.
The wave packet parameters and their equations of motion only enter as inter-
mediaries.

A common feature in both nuclear and molecular wave packet theory is
the fact that wave packets are being used with a dual purpose. A
first application is the definition of a suitable basis in Hilbert space
(or in a sub-space). Here the wave packet description is essentially a
static one. The aim is the diagonalization of the Hamiltonian to obtain
the stationary states of the system. The time dependence (dynamics) enters
only to establish a method of picking a discretized basis out of the continuous manifold of wave packets. A second use of wave packets is as a tool
for the physical interpretation of certain types of motion. Here the wave
packet is considered as a non-stationary initial state and its time evolution
is interpreted to display the dynamics of the system.

One of the crucial problems occuring both in molecular and nuclear
applications is that of the long-time dynamics. Only in the simplest cases
can the time-dependent Schrödinger equation be solved. In most applications
one relies on an approximate time-evolution. In nuclear physics it is
common to use the time-dependent Hartree-Fock method or more generally, the
time-dependent variational principle. In molecular physics the foundations
of approximate dynamics seems to be less systematic. In any case there is no
guarantee that the wave packet, propagating according to these approximate
dynamics, will for long times remain close to its exact time evolution.
While some investigators would discourage the use of time-dependent wave
packets altogether, some numerical calculations indicate that approximate
dynamics remain meaningful. A closer study is called for.

Another question of common concern is that of the quantization rules to
be applied to the classical equations of motion that govern the wave packet
parameters. One possibility well-known in nuclear physics is the Generator
Coordinate Method. In this procedure the wave-packet is considered as the
intrinsic function and   time as the Generator Coordinate. The method is,
in fact nothing else but the diagonalization in a subspace of wave packets.
Another possibility, used frequently in molecular physics, is to look for
the energy eigenvalues on the Fourier Transform of the auto-correlation
function. Furthermore there are those quantization methods which rely on
a Bohr-Sommerfeld-like rule applied to phase-space trajectories. From the
discussions it can be inferred that the multidimensional cases of experimental interest will present serious computational problems. A related problem
is the elimination of spurious eigenvalues. A comparative study of different
quantization methods would certainly yield valuable information.

Besides the rather specific aspects of wave packet theory mentioned above, several other topics, of relevance in a broader context, came up during the discussions. One of these was the relationship between classical and Quantum Physics. The problem was posed as to how much of the quantal properties of a system can be calculated from the knowledge of the classical solutions of the corresponding classiccal system (if any). A synthesis of practical results obtained by the wave packet applications, should be useful for future work on this question.

A second major theme underlying the discussions was the relation between linear and nonlinear theories. Indeed the wave packet description leads one from the linear Schrödinger equation to non-linear Hamilton equations. Although quantum chaos was not a subject for the conference it came up on several occasions. The connection between spectral distributions and wave packet trajectories may be another point of further study.

On the methodological side it was suggested several times that, in order to make progress in realistic systems, one should concentrate on reduced descriptions. This means that the system is separated in two parts, one part to be treated by classical mechanics and the other by quantum mechanics. An important aspect of this situation is the energy transfer between the parts. Various examples were mentioned: solvents and reactive particles in molecular dynamics, collective and particle degrees of freedom in nuclei, electronic and nuclear motion in molecules, radiation field and matter, heavy ion collisions, etc....

As a final conclusion it may be stated that there is a general agreement among the participants of the meeting that wave packets dynamics constitutes a valuable theoretical instrument both in molecular and nuclear physics. It is of interest for both fields of research to develop further the techniques of wave packet propagation. Especially the comparison of various approximation schemes for the long time dynamics appears to be a fruitful open question. Also the comparison of various quantization methods is worth a thourough investigation. To promote further advancement in the theory of wave packet dynamics a test problem should be formulated which shares features of nuclear and molecular theory. The application of known techniques from both fields should give insight in the reliability of wave packet dynamics as a theoretical tool.

LIST OF PARTICIPANTS

ARICKX Frans, Dienst Teoretische en Wiskundige Natuurkunde, Rijksuniversitair
    Centrum Antwerpen, Groenenborgerlaan 171, 2020 Antwerpen, Belgium.
ARVIEU R.,Institut des Sciences Nucléaires, Université de Grenoble,
    53 Av. des Martyrs, F-38026 Grenoble, France.
BERGER Jean François, Commissariat à l'énergie Atomique, Centre d'études de
    Bruyères le Chatel, B.P. 12, 91680 Bruyères Le Chatel, France.
BRICKMANN Jürgen, Physikalische Chemie, Technische Hochschule Darmstadt,
    Petersenstr. 20, 6100 Darmstadt, D.B.R..
BROECKHOVE Jan, Dienst Teoretische en Wiskundige Natuurkunde, Rijksuniversitair
    Centrum Antwerpen, Groenenborgerlaan 171, 2020 Antwerpen, Belgium.
CERJAN Charles, Lawrence Livermore National Laboratory, University of California,
    P.O.B. 808, Livermore, CA 94550, U.S.A..
COALSON Rob, Los Alamos National Laboratory, Los Alamos, New Mexico 87545, U.S.A..
CRIBB Peter, Physikalische Chemie, Technische Hochschule Darmstadt,
    Petersenstr. 20, 6100 Darmstadt, D.B.R..
DEUMENS Erik, Quantum Theory Project, Williamson Hall 365, University of
    Florida, Gainesville, Florida 32611, U.S.A..
DREIZLER Reiner, Institut für Theoretische Physik, Universität Frankfurt,
    Robert-Mayer Strasse 8-10, Postfach 11 19 32, 6000 Frankfurt/Main, D.B.R..
GIRAUD Bertrand, C.E.N. de Saclay, Service de Physique Théorique, B.P.2,
    91190 Gif-sur-Yvette, France.
GRIFFIN James, University of Maryland, Department of Physics and Astronomy,
    College Park, MD 20742, U.S.A..
HAHN Yukap, Department of Physics, The University of Connecticut, Storrs,
    Connecticut 06268, U.S.A..
HELLER Eric, Department of Chemistry and Physics, University of Washington,
    Seattle WA 98195, U.S.A..
JOACHAIN Charles, Physique Théorique, U.L.B., Campus de la Plaine, Boulevard
    du Triomphe, 1050 Bruxelles, Belgium.
KAN Kit Keung, JAYCOR, 205 South Whiting Street, Alexandria CA 22304, U.S.A..
KESTELOOT Eddy, Dienst Teoretische en Wiskundige Natuurkunde, Rijksuniversitair
    Centrum Antwerpen, Groenenborgerlaan 171, 2020 Antwerpen, Belgium.
KOHL Herfried, Institut für Theoretische Physik, Universität Frankfurt,
    Robert-Mayer Strasse 8-10, Postfach 11 19 32, 6000 Frankfurt/Main, D.B.R..
LATHOUWERS Luc, Dienst Teoretische en Wiskundige Natuurkunde, Rijksuniversitair
    Centrum Antwerpen, Groenenborgerlaaan 171, 2020 Antwerpen, Belgium.
LE FORESTIER Claude, Laboratoire de Chimie Théorique, Université de Paris-Sud,
    91405 Orsay, France.
MANKOC-BORSTNIK Norma, Department of Physics, E. Kardelj University, P.O.B. 543,
    Jadranska 19, 61001 Ljubljana, Yugoslavia.
MEYER Hans-Dieter, Theoretische Chemie, Universität Heidelberg, Im Neuenheimer
    Feld 253, D 6900 Heidelberg, D.B.R..
MOISEYEV Nimrod, Chemistry Department, Technion Israel, Institute of Technology,
    32000 Haifa, Israel.
MOSER Carl, C.E.C.A.M., Université Paris-Sud, Batiment 506, 91405 Orsay Cedex,
    France.
NIETO Michael Martin, Los Alamos Nat. Lab., Los Alamos, New Mexico 87545, U.S.A..
PLOSZAJCZAK Marek, Niels Bohr Institute, Blegdamsvej 17, 2100 Copenhagen, Denmark.
REIMERS Jeff, Theoretical Chemistry Department F11, University of Sidney
    NSW 2006, Australia.
SCHUCK Peter, Institut des Sciences Nucléaires, 53 Avenue des Martyrs,
    38026 Grenoble Cedex, France.
SINGER Konrad, Royal Holloway College, University of London, Department of
    Chemistry, Egham Hill, Surrey TW20OEX, Egham 35351, G.B..
SKODJE Rex, Department of Chemistry, Campus Box 215, University of Colorado,
    Boulder, Colorado 80309, U.S.A..
VAN LEUVEN Piet, Dienst Teoretische en Wiskundige Natuurkunde, Rijkuniversitair
    Centrum Antwerpen, Groenenborgerlaan 171, 2020 Antwerpen, Belgium.
ZIMMERMAN Theo, Theoretische Chemie, Universität Heidelberg, In Neuenheimer
    Feld 253, D 6900 Heidelberg, D.B.R..